Biology
Reviewing the Essentials

Caroline B. Hopkins
Phone #'s (800) 522-2313
(704) 522-7178
Email: chopkins97@aol.com

PBD, Inc.
3280 Summit Ridge Pkwy. Suite 100
Duluth, GA 30096
Phone # (800) 531-0071
Fax # (770) 442-9742
Email: Schoolbk@pbd.com

Biology
Reviewing the Essentials

Dean Medley
Teacher on Special Assignment
Office of Professional Induction and Success
Former Biology Teacher
Guilford County School System, Greensboro, NC

Former Assistant to the Director
UNC Mathematics and Science Education Network
University of North Carolina, Chapel Hill, NC

AMSCO

AMSCO SCHOOL PUBLICATIONS, INC.
315 Hudson Street / New York, N.Y. 10013

Please visit our Web site at:

www.amscopub.com

Cover photo: Green violet-ear hummingbird, in cloud forest, Costa Rica, by Michael Fogden

When ordering this book, please specify:
either R 416 P *or* BIOLOGY: REVIEWING THE ESSENTIALS

ISBN 0-87720-062-9

Printed in the United States of America

6 7 8 9   02 01

PREFACE

YOU MIGHT be surprised to learn that a one-year high school science course introduces you to about the same number of vocabulary words as a first-year course in a foreign language. For many students, science vocabulary is as confusing and difficult to learn as the vocabulary needed to understand a foreign language.

However, the importance of science in everyone's lives has increased dramatically. It is important to be aware of many of the terms and concepts presented in your science courses in order for you to make wise decisions in the future. A quick look at a daily newspaper will prove the increasing importance of science in modern society. Many of the unanswered scientific questions that face us may be solved by you and your classmates. The *Biology: Reviewing the Essentials* you hold in your hands will help support you in your study of biology—quite possibly the most important and relevant science course you will study.

This book was designed and written with you in mind. It is here to help you master the subject matter your teacher will present this year. The scope of this book permits it to be used in many different learning situations. The language used recognizes the diverse vocabulary levels of students. Chapters are arranged to follow a standard one-year course in biology, but the order in which you read them can be changed to correspond with the textbook you are now using.

The outstanding features of this book are as follows:

1. Presentation: The content of this review text has been organized into twelve units. Each unit is divided into a series of short chapters written in a simple narrative style that makes the content more easily understood. Important words to remember are **boldfaced** to make them stand out. These words are defined in the text where they appear for the first time. It might help you remember these words if you copied them into your notebook along with their definitions. Then you can use this list when you review for an exam. An index allows you to locate key words and topics quickly.

2. Illustration program: When appropriate, line illustrations are integrated in the text. These illustrations present a different way for you to learn. They are drawn by artists who are skilled in presenting scientific material in an accurate and interesting manner. You may want to copy some of these illustrations in your notebook. It really doesn't matter if you are a skilled artist—these drawings are meant to help you learn and organize the biology information you will be taught this year.

Charts, graphs, and tables are also included throughout this text. These graphics present another way to process information that many students find useful. Information in the charts and tables is usually presented in a brief way. The "key words" included help students organize and synthesize the data presented. You may wish to copy the charts and tables into your notebook. Your teacher may wish to write the charts and tables on the chalkboard. The illustrations in this review text are an integral part of the program—you should use them as additional study aids.

3. Questions: The questions presented at the end of each chapter help you master the material presented. If you answer a question incorrectly, you should reread the section of the text that covered that material. The questions in this review text are not meant to be tricky. They are written to test your understanding of the material, and to help you recognize gaps in your knowledge so that you can review that topic. A variety of different question types are included, such as multiple-choice, matching, fill-in, essay, and portfolio questions. These different types of questions will help you review as you prepare for exams.

Biology: Reviewing the Essentials will prove to be an effective aid for students in mastering the material presented in a one-year course in biology. The biological world surrounds you, and includes you. This brief review text will help you explore the world of living things—a world in which you play an important part.

CONTENTS

UNIT SIX
CLASSIFICATION OF LIFE

UNIT SEVEN
THE MICROSCOPIC WORLD

UNIT EIGHT
MANY-CELLED ORGANISMS

UNIT NINE
BIOLOGY OF HUMANS

UNIT TEN

BEHAVIOR OF LIVING THINGS

UNIT ELEVEN

ECOLOGY: WEBS OF LIFE

UNIT TWELVE

FUTURE HORIZONS IN BIOLOGY

UNIT ONE
THE NATURE OF SCIENCE

Chapter 1. Scientific Method and Measurement

Chapter 2. Science and Technology: Benefits and Limitations

Chapter 1
Scientific Method and Measurement

EACH DAY WE ask questions, answer questions, and try to solve problems. We use past experiences, education, common sense, and desire to help us find answers to questions that occur.

Scientists are in the business of answering questions. Each time a scientist solves a problem, everyone benefits because the solution becomes part of the body of knowledge available to all people. However, an interesting thing often happens in the process of finding answers. Scientists often generate still more questions about the natural world that need to be answered. In many ways, the work of science is a cycle—answers generate more questions, which require another search for answers, which often generate even more questions.

The process of systematically searching for answers probably began thousands of years ago, when people became more aware of their surroundings and the many changes that occurred in their world. They made observations and began to use these observations to predict future events. Our search for answers has had many practical effects and allows people continually to improve the quality of their lives.

THE SCIENTIFIC METHOD

Today, scientists follow a definite, logical process when trying to solve a problem. They do this because it makes their efforts more consistent, easier to analyze, and easier to communicate with other scientists. This process has a series of specific steps and is called the **scientific method.** The Italian scientist Galileo is credited with being the first person to use the scientific method.

The first step in the scientific method is to **state the problem** or question that needs to be solved. Sometimes asking a question correctly is more difficult than finding the correct answer. After stating the problem, a scientist (in this case, you) must **collect information** about the subject. This information may be found in books, magazines, journals, and videos, and by talking to knowledgeable people. Today, information is also available on computer databases. By gathering as much background information as possible, you are better able to design an experiment to solve a problem and to evaluate the progress of your experiment and the results you obtain.

You must now make a guess about how you think your experiment will work out. This guess is called **forming your hypothesis.** In part, your hypothesis is a reflection of your background knowledge. Remember that a hypothesis is a proposed answer for the question you want to solve. It can be thought of as a short-term, or temporary, explanation based on your early observations.

To find out if your hypothesis is correct, you must **perform an experiment.** Every experiment must have both an experimental group and a control group. The **experimental group** will contain a variable. A **variable** is one factor in the experiment that is changed. In any experiment, you can have only one variable. All other factors must remain constant. It is the variable that may produce a change in the results of the experiment. The final results may support—or not support—your hypothesis. The **control group** is used as a point of comparison for the experimental group. It illustrates what happens if nothing is changed (that is, if there is no variable). Experiments should be performed more than once to make sure the results that you obtain are accurate.

The following example will provide some understanding of how to set up an experiment. Suppose your hypothesis is that a special chemical, found in plants, affects how quickly a stem grows. To test this hypothesis, you need to set up two identical flowerpots with equal amounts of soil. Then you plant several seeds in each pot and label one pot "Treated" and the other pot

"Untreated." The treated pot contains your experimental group; the untreated pot contains your control group. After the seeds sprout, and are at the same stage of growth, you measure the seedlings in each pot. Then you spray the plants in the treated pot with the chemical you are testing. Of course, you leave the control group untreated. You expose both pots to the same amount of sunlight and give them the same amount of water. You measure the plants several times over a period of several weeks. By comparing the heights of the plants in the two pots, you determine what effect, if any, the chemical has on the growth of the plants.

Next, you **record the observations** that you made throughout the experiment. These observations are recorded as **data**. The data describe what happened in the experiment. Data may be recorded in the form of graphs, tables, drawings, photographs, or words.

After analyzing the data you collected, you must **form a conclusion** about your results. Your conclusion may or may not support your hypothesis. If the hypothesis is not supported by the results of your experiment, you may decide to perform the

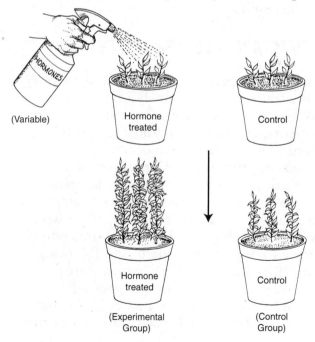

Figure 1-1. All experiments need a control. Why?

experiment again, or you may design a new experiment to learn more about the problem. A hypothesis that is supported many times through experimentation may become a **scientific theory.** Unlike a theory proposed by a television detective, which is at best a guess or a hunch, a scientific theory is a widely accepted explanation for a natural phenomenon. But even a scientific theory can be changed if new, contrary evidence is gathered.

The last step in the scientific method is to **share your results** with others. Sharing results is one of the most important steps in a scientist's work. It helps other scientists avoid duplicating work, and this allows research to proceed more quickly. For example, the sharing of information is extremely important in our fight against life-threatening diseases. For many years, because of political and other considerations, scientists in some countries were unwilling or unable to share information with scientists working in other countries. Who knows how much information was lost and efforts duplicated because of this lack of communication? There is another important consideration in sharing information. Sharing the results of experimental work also allows other scientists to check the accuracy of the results.

SCIENTIFIC MEASUREMENTS

When scientists perform important experiments, communication and accuracy are extremely important. So when scientists make observations during an experiment, they must take **measurements**. Measurements allow scientists to make specific observations of any changes that occur.

Scientists use the **International System of Measurement** (SI). This is a decimal system based on multiples of 10. Because this system is used by all scientists, no matter what language they speak, they are able to share their results with other scientists anywhere in the world and be understood. Scientific measurements are usually made of the following: length (distance from one point to another); temperature (warmth or coldness of a substance); volume (amount of space an object occupies); and mass (amount of matter in an object). Table 1-1 illustrates the most commonly used units of measurement in the SI system.

Table 1-1.   Common Units of Measurement

Measurement	Unit	Symbol
Length	meter	m
Mass	gram	g
Volume	liter	l
Temperature	degree Celsius	°C

By adding prefixes to the unit names (meter, gram, and liter), larger or smaller units can be described. The following table includes some of the prefixes most commonly used in scientific work.

Table 1–2.   Prefixes Used With Numbers

Prefix	Meaning
mega	1,000,000x
kilo	1,000x
hecto	100x
deka	10x
deci	1/10th
centi	1/100th
milli	1/1000th
micro	1/1,000,000th
nano	1/1,000,000,000th

QUESTIONS

Multiple Choice

1. Which of the following scientists is credited with being the founder of the scientific method?   *a.* Mendel   *b.* Smith   *c.* Galileo   *d.* Darwin

2. A hypothesis that has been supported many times is called a   *a.* conclusion   *b.* control   *c.* theory   *d.* law.

3. The first step in the scientific method is to   *a.* experiment   *b.* observe   *c.* form a hypothesis   *d.* state the problem.

4. The experimental group always contains a   *a.* hypothesis
   b. variable   *c.* theory   *d.* control.
5. Which of the following is the proper unit for the measurement of mass?   *a.* meter   *b.* Celsius   *c.* liter
   d. gram

Matching

6. The problem of buying a car using the scientific method.

   a. I think an economy   _____(1) define the problem
   car is my best buy.   _____(2) gather information
   b. This make of econ-   _____(3) hypothesis
   omy car is the car I
   am going to buy.   _____(4) experiment
   c. I am spending a great   _____(5) form a conclusion
   deal of money keeping   _____(6) share the information
   this old car running. It
   is time to buy a new
   car.
   d. I will test-drive
   several cars.
   e. I will read car magazines to find the latest
   information.
   f. I will take my friends
   for a ride in my new
   car.

7. Arrange these steps in the scientific method in the proper
   order.
   a. experimentation   (1) _____
   b. conclusion   (2) _____
   c. hypothesis   (3) _____
   d. gathering information   (4) _____
   e. defining the problem   (5) _____
   f. sharing the new knowledge   (6) _____

Free Response

8. What unit would you use to measure the following? Why?
    a. your height
    b. the mass of a pencil
    c. the amount of water in a glass
9. Why do scientists use measurements when recording data?
10. List four places a scientist could go to find information about a problem.
11. What is the best unit of measurement to use with each example?
    a. the mass of a penny—kilogram or milligram
    b. volume of a soft drink—liter or kiloliter
    c. your height—centimeter or kilometer
    d. the length of a sheet of paper—centimeter or kilometer
    e. distance from New York City to Miami—centimeter or kilometer
    f. length of an eyelash—millimeter or kilometer
12. Which is larger?
    a. millimeter or centimeter
    b. kilogram or gram
    c. dekaliter or deciliter
    d. centimeter or megameter
    e. microliter or liter

Portfolio

The great fictional detective Sherlock Holmes used the scientific method to solve many mysteries. Read one of the Sherlock Holmes stories and write a brief report that describes Holmes's scientific approach to crime.

Chapter 2

Science and Technology: Benefits and Limitations

SCIENCE IS A GROWING, changing body of information that has been collected over many years. The word **science** comes from a Latin word meaning "to know." Science concerns itself with things that are experienced, observed, and measurable. Today, science plays an important role in everyone's life.

People who work in science are called **scientists.** Scientists try to find the reasons or explanations for natural phenomena. They ask questions and seek observable answers. Scientists must try to remain objective and free of bias in their work. They cannot make value judgments in their study. The knowledge gained from their studies can, however, provide information that will help people make better, more practical decisions.

Many nonscientists feel that science can and should provide solutions to all the problems of human life. Others see science as posing a threat to normal human existence. Some people even have the perception that scientists are often out of touch with reality, working at bizarre tasks in mysterious, equipment-filled laboratories. Scientists are actually people who are curious about the natural world and who want to explore it further.

LIMITATIONS TO SCIENTIFIC WORK

Over the years, many problems have limited scientific progress. Some of these limitations continue to slow the search

Figure 2-1. Everyone can contribute to scientific knowledge.

for answers to the many problems facing us today. For example, scientists may be limited by a lack of modern equipment. Without proper equipment and facilities, scientists may not be able to answer some of the many questions facing them. A lack of funding is another major concern for many scientists. This is not to suggest that all problems could be answered if enough money were provided for research, but increased funding could expand the amount of research that scientists can perform. Because the government provides funds for basic and applied research, politics often plays an important role in scientific research.

Science has also been limited by society's perceptions. People are sometimes reluctant to accept scientific ideas, particularly if these ideas conflict with deeply held beliefs. Medicine has been especially affected by people's reluctance to try new treatments and procedures. Society must also place a value on scientific work. Many new discoveries have the potential to harm, as well as help. New inventions may affect the environment. For example, modern methods of transportation use up Earth's fossil fuel resources. The burning of fossil fuels may also cause new problems, such as air pollution, which then have to be solved.

SCIENCE AND TECHNOLOGY

Scientists use a variety of techniques and equipment to assist them in carrying out an experiment. Science and technology work together in the areas of discovery and development. **Technology** is a method or process for handling a problem and may be thought of as a way of applying science to everyday problems.

Technology has also enlarged the boundaries of scientific work. For example, the electron microscope provides us with views of a world so small that it could only be imagined a hundred years ago. Today, we can see the extremely tiny structures that are the parts of microscopic cells. Scientists can even operate on these cell parts, which are made visible through these techniques of magnification.

Other tools enable scientists to peer into the human body without performing surgery. MRI (magnetic resonance imaging) can make pictures of organs inside the body. Previously, only X rays provided views inside the body, but X rays can show only dense materials such as bone. MRI can show structures of less dense material, such as the brain and other internal organs. Unlike X-ray images, MRI does not expose the body to radiation.

Computers allow us to store and analyze vast amounts of data. The speed at which computers perform their tasks has saved time and effort in completing experiments. Discoveries and the results of experiments can also be transmitted to scientists around the world by computers. In the past, these results were only printed in journals and books, a process that was much more costly and time-consuming.

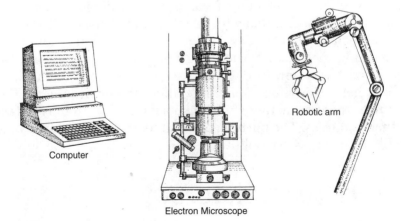

Computer

Electron Microscope

Robotic arm

Figure 2-2. Technology has enhanced our ability to study the natural world.

The new area of biotechnology has revolutionized the field of genetics—the study of traits that are passed from one generation to the next. Since certain human traits are unique to a specific individual, biotechnology has also changed the ways crimes are investigated and the techniques used to identify an individual involved in a crime.

Other types of technology have increased our efficiency at performing experimental procedures. Robots now serve as mechanical aids and have many applications. Robots are used to make cars, and they are also used to manipulate materials in the laboratory. They are especially valuable in manipulating substances that may pose a danger to human researchers.

The world of technology changes almost daily. Scientists use technology to discover facts that will help us understand our world. However, the world of science is so huge that it is divided into different special fields of study. Many scientists specialize in a narrow topic of study. However, scientists who work in a specific area often need to combine their efforts in order to fully understand a particular area. The following table lists some of the important fields of scientific study.

Table 2–1.   Fields of Study in Biology

Biology	The study of living things
Chemistry	The study of matter and the changes that occur in matter
Physics	The study of how energy and matter are related
Astronomy	The study of stars, planets, and galaxies
Geology	The study of rocks and minerals and the various processes that occur within Earth
Oceanography	An application of several sciences to the study of the world's oceans

QUESTIONS

Multiple Choice

1. The application of scientific discoveries to the solving of everyday problems is called   *a*. discovery   *b*. technology   *c*. biology   *d*. observation.

2. A study of our solar system would be part of the science of   *a*. chemistry   *b*. astronomy   *c*. biology   *d*. oceanography.

3. The job of a scientist is to   *a.* ask questions
   b. be objective   *c.* seek answers   *d.* all of these.

4. The word *science* comes from a Latin word meaning   *a.* to research   *b.* to experiment   *c.* to observe   *d.* to know.

5. Which of the following would be most likely to study a reaction in a test tube?   *a.* a physicist   *b.* a geologist *c.* a chemist   *d.* an astronomer.

Matching

6. physics
7. chemistry
8. biology
9. astronomy
10. geology

a. study of living things
b. study of planets and galaxies
c. study of matter and its changes
d. study of energy and matter
e. study of minerals and rocks

Free Response

11. Which field of science would study the following? (There may be more than one correct answer. Be able to support your answer.)

    a. Collect and study insects   _____

    b. Look through a telescope   _____

    c. Time reactions in a test tube   _____

    d. Take water samples   _____

12. What are some characteristics that make a good scientist?

13. What are three limitations to scientific research?

14. How does science differ from other subjects you study?

Portfolio

Use your imagination to design a nonpolluting car that does not use fossil fuel. You can make a model of this car, or a drawing. You can also describe the car in words. Share your ideas with your class.

UNIT TWO
BIOLOGY: THE SCIENCE OF LIFE

Chapter 3

The Science of Biology

MILLIONS OF DIFFERENT types of organisms exist on Earth today. The study of these living things is called **biology**. The word biology comes from the Greek word *bios,* meaning "life," and *logia*, which means "the study of." The science of biology allows us to better understand ourselves and all the other living things that share our planet.

Biology, like all sciences, has its roots in the curiosity of humans and was originally based on casual observations of the natural world. Later, simple experiments were performed to help answer questions about living things. Today, the study of biology forms the basis of many professions, such as medicine and agriculture. Biologists may work in a laboratory environment or in "the field," surrounded by the living organisms they are studying.

CLASSIFYING LIVING THINGS

Aristotle, a Greek philosopher, is considered to be "the founder of biology." Aristotle made wonderful observations of the world and even developed a system for classifying living things. Classifying organisms makes it easier to study them.

The system of classification we use today was developed by **Carolus Linnaeus** in the 1700s. Linnaeus is known as the founder of modern classification. Linnaeus classified organisms as either plants or animals. He developed a two-word naming system, which is known as **binomial nomenclature.**

In 1859, **Charles Darwin** proposed his theory of **evolution.** This theory shows the historical relationships that exist among living things. It has had a profound effect on the development of modern systems of classification, which recognize the evolutionary relationships of organisms.

BIOLOGISTS AT WORK

The following is a brief list of people who made contributions to the body of biological knowledge.

Anton van Leeuwenhoek, a Dutch lens maker, is credited with making the first microscope. Leeuwenhoek was the first person to observe microscopic organisms.

Robert Hooke was an Englishman who observed cork—the protective layer of cells produced by trees—under a microscope. He noticed that the cork sample was divided into small chambers. He called these chambers cells. Supposedly, they reminded him of the cells, or rooms, in a monastery. Today, the word cell describes the smallest structural unit of all living things.

Carolus Linnaeus was a Swedish botanist who developed the system for naming organisms used by biologists. Known as binomial nomenclature, this system links a unique genus and species name to every type of living thing. These two names are used by scientists all over the world to describe particular organisms. Before this system was developed, people used many different names for the same organism. Linnaeus's work brought order to what was previously a chaotic situation.

Charles Darwin was the British scientist who developed the idea of **natural selection,** which led to his famous theory of evolution. Darwin's theory recognizes that life has existed on Earth for a very long time, and has changed over time to produce the many types of living things we observe today.

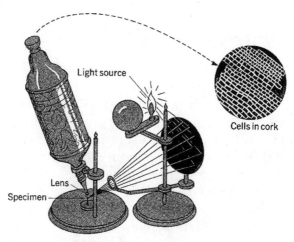

Figure 3-1. Hooke's microscope.

Gregor Mendel was an Austrian monk whose work formed the basis for the modern science of genetics. Mendel's work with pea plants enabled him to develop the basic laws that are used to explain the inheritance of traits.

Matthias Schleiden, a botanist, and **Theodor Schwann,** a zoologist, made observations that led to the "cell theory." The cell theory states that: All living things are made up of cells; cells are the basic unit of structure and function of all living things; and all cells come from preexisting cells.

James Watson, an American biologist, and **Francis Crick,** a British physicist, discovered the structure of DNA. DNA is the nucleic acid that stores information needed for all cellular activities. Their work was based, in part, on the brilliant X rays of DNA taken by biochemists Rosalind Franklin and Maurice Wilkins.

Rachel Carson, an American writer and biologist, warned of the danger of the increased use of pesticides and the damage it was doing to nature. This brilliant scientific writer attracted a large audience. Her book *Silent Spring* made Americans aware of ecology and the concept that all living things are important, because their lives are interconnected in many ways.

Jacques-Yves Cousteau, a French ocean explorer, is the coinventor of the aqualung. This device revolutionized underwater exploration by making it easier for people to explore the world of ocean life. He introduced the general public to life in the sea through his books and films.

TOOLS OF THE BIOLOGIST

Today's biologists have many tools to make their search for answers more successful. Probably the most important tool used by the biologist is the **microscope.** The microscope allows biologists to see small objects in great detail. Anton van Leeuwenhoek built simple microscopes that were able to magnify up to 300 times. The compound-light microscope used today has a much greater power of magnification (up to almost 1000 times), and better resolution (clarity or sharpness) than early microscopes.

In the 1930s, scientists developed the first **electron microscope.** This type of microscope used beams of electrons, instead of light, to make an image. Today, there are two types of electron microscopes, the **transmission electron microscope** (TEM) and the **scanning electron microscope** (SEM). In the TEM, electrons actually pass through the object being viewed. The biologist sees a thin, flat view of the internal structures of a specimen. The SEM gives the biologist a surface view of a specimen by coating the specimen with metal, causing the electrons to bounce off the surface. Special detectors pick up the electrons and convert them into an image on a television screen.

Computers have also increased our knowledge by storing and processing great quantities of data.

There are many fields of biological study. The variety of topics in biology makes it an interesting subject to study. Because of the diversity and complexity of living things, biologists usually limit their areas of interest. Below is a list of a few specific areas of study in biology.

The science of biology will affect our future in many profound and interesting ways. Biology will provide a basis of knowledge that will allow us to make well-informed choices in the future. It will help us deal intelligently with many routine concerns, such as nutrition and health. Biology will allow us to learn more about ourselves and the world we live in, and answer questions that may improve the quality of human life. As a possible career path, biology offers interesting opportunities for students.

Table 3–1.   Fields of Study

Anatomy	The study of the structure of the body
Botany	The study of plants
Cytology	The study of cells
Ecology	The study of the interaction of organisms with their environment
Embryology	The study of the early developmental stages of an organism's life
Genetics	The study of how characteristics are passed from parents to offspring
Physiology	The study of the internal functions of organisms
Marine biology	The study of living things in the ocean
Microbiology	The study of microscopic life
Taxonomy	The classification of living things
Zoology	The study of animals

QUESTIONS

Multiple Choice

1. The modern system of classification was developed by
   a. Darwin   *b.* Linnaeus   *c.* Hooke   *d.* Carson.
2. The "founder" of biology was   *a.* Hooke   *b.* Darwin
   c. Leeuwenhoek   *d.* Aristotle.
3. Mendel became famous for his study of   *a.* zoology
   b. heredity   *c.* cytology   *d.* bacteriology.
4. Watson and Crick discovered the structure of   *a.* proteins
   b. DNA   *c.* RNA   *d.* carbohydrates.
5. Learning the bones of the human body would be part of
   the study of   *a.* cytology   *b.* physiology   *c.* embryology
   d. anatomy.
6. Which of the following microscopes allows you to see a
   surface view of an object?   *a.* simple microscope
   b. TEM   *c.* SEM   *d.* compound-light microscope

Matching

7.	zoology	*a.*	plants
8.	anatomy	*b.*	animals
9.	physiology	*c.*	structure of the body
10.	ecology	*d.*	organisms and their environment
11.	botany	*e.*	functions of the body

Fill In

12. Biology comes from the words _____, meaning "life," and
    _____, meaning "the study of."
13. What are some of the ways biology affects your life?
14. Why are there many areas of biology to study?

Portfolio

In the future, exobiology—the study of life away from planet
Earth—may be an important area of study. Write a short science fiction story about a trip to another planet in another
solar system that supports life. You may want to include drawings of several kinds of organisms that live there.

Chapter 4

Characteristics of Life

IT IS USUALLY easy to recognize life, but it is often much harder to define it. **All living things are made of cells.** Some organisms are **unicellular** and consist of only a single cell. Other organisms are **multicellular** and are composed of many cells. To determine whether an object is living or nonliving, biologists have agreed on several characteristics that define living things. They are referred to as **life processes** or activities. These life processes include such activities as growth, metabolism, movement, and reproduction. Living things also react, or respond, to their environment. The ability to respond to an environmental stimulus is called **irritability** (no, this word does not mean cranky in this case).

LIFE PROCESSES

Living things grow. **Growth** is an increase in size. Most organisms also go through a series of changes called **development.** The beginning form of an organism may not resemble its adult form. For example, a tadpole does not look the same as an adult frog. Growth in multicellular organisms is due to an increase in the number of cells. Humans begin life as a single cell. However, when they are born they are made up of trillions of cells. When they are adults, they consist of even more cells.

Metabolism refers to the chemical activities that are needed for life. Ingestion, digestion, respiration, and excretion are the processes of metabolism. **Ingestion** is taking in food. The process of breaking down food into simpler substances is called **digestion.** The breaking down of food particles to release energy is called **respiration.** For biologists, respiration has two meanings. Respiration occurs at the cellular level when food is broken down to release energy. Respiration is also the process of taking in oxygen and giving off carbon dioxide as a waste gas. In humans, breathing and respiration are often used to refer to the same process. **Excretion** is getting rid of wastes. Excretion usually refers to the removal of solid and liquid metabolic wastes that are produced during respiration.

Catabolism and anabolism are two processes in living things that are also involved in metabolism. **Catabolism** is the breakdown of complex substances into simpler substances. **Anabolism** is the formation of complex substances from simpler substances.

Reproduction is the process of producing new organisms of the same kind. Reproduction of living things can occur asexually, requiring only one parent, or sexually, requiring two parents. Organisms that consist of a single cell reproduce asexually by dividing. Organisms that reproduce sexually contain genetic material contributed from each parent. If a group of living things does not reproduce, extinction of that group occurs.

Living things react to changes in their environment. A **response** is a reaction to a change. Responding to a change in the environment may increase an organism's ability to survive.

Organisms must be able to get and use energy in order to survive. **Energy** is needed to carry out all cellular processes. For example, organisms use energy from food to grow, develop, and reproduce.

Energy flows through individual animals, communities, and the environment. It is passed from one organism to another organism, usually in the form of food. For example, a zebra eats the energy from the sun that is contained in the cells of a plant. A lion then eats the zebra and gets its energy. In this way, energy moves through living systems from one organism to another. The sun is the ultimate energy source for most of the organisms that live on Earth, although there are exceptions. Some bacteria, for example, are able to use the energy trapped in chemical compounds rather than the energy from the sun as plants do.

Living things are **highly ordered.** A tree grows into a form typical of its species. All humans have the same general form, although there are differences in size. The chemical reactions that occur in living things do not occur randomly. The chemicals that make up living organisms are, in general, more complex than the chemicals found in nonliving things, such as rocks. All living things are complex. All are composed of small units of life called cells. Cells are able to carry out all the life processes that insure their survival.

LEVELS OF ORGANIZATION

Scientists have recognized **levels of organization** to illustrate the structure of living things. **Cells,** since they are the smallest units of life, are the basis of the organization. Similar

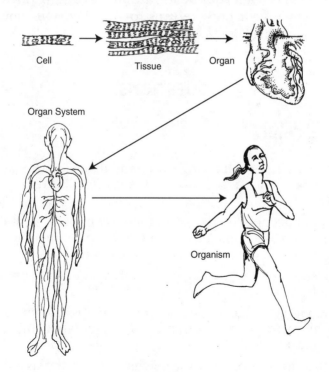

Figure 4-1. Levels of organization.

cells that work together form **tissues.** Muscle, blood, bone, and nerves are all tissues. Each tissue has a specific function. This is called division of labor. Two or more different tissues working together form an **organ.** The heart, lungs, and kidneys are organs. Organs work together to carry out a definite function in a **system.** In the human body, there are digestive, respiratory, circulatory, excretory, reproductive, and nervous systems. When groups of systems work together, the result is an **organism.** Each system must function properly to maintain the welfare of the whole organism.

All body systems work together to maintain a stable condition. This stable condition is called **homeostasis.** During the day, your body maintains a constant temperature, a regular heartbeat, and a steady breathing rate. At night, while you sleep, your body also maintains a constant state or homeostasis. This does not mean that your body does not react to changes. If you run around a track 10 times, your temperature remains fairly constant even though your heart rate and breathing rate change. However, when your body returns to a resting state, your heart rate and breathing rate slow down.

QUESTIONS

Multiple Choice

1. Life activities such as ingestion and digestion are part of the process of   *a.* growth   *b.* respiration   *c.* response   *d.* metabolism.

2. Growth occurs due to   *a.* an increase in the size of cells   *b.* an increase in the number of cells   *c.* an increase in metabolism   *d.* a lowering in the rate of cell division.

3. The smallest, most basic unit of life is the   *a.* element   *b.* cell   *c.* system   *d.* tissue.

4. When a tissue has a specific function, it is known as   *a.* unlimited   *b.* division of labor   *c.* metabolism   *d.* noncellular.

5. In the levels of organization, what follows the tissue level?   *a.* cells   *b.* organisms   *c.* organs   *d.* systems.

Matching

6. growth
7. cell
8. catabolism
9. anabolism

10. reproduction

a. forming complex substances
b. the smallest unit of life
c. breaking down complex substances
d. producing organisms of the same kind

e. an increase in size

Fill In

11. One-celled organisms are called _____.
12. _____ reproduction requires only one parent.
13. Responding to the environment increases an organism's ability to _____.

Portfolio

Are viruses alive? Scientists disagree. Find out about viruses. Write a short newspaper article that describes some of the difficulties in determining whether viruses are alive.

Chapter 5

Abiogenesis vs. Biogenesis

THE HISTORY OF biology is characterized by change, as yesterday's "facts" often become today's fiction. For example, in the past, many people believed that living organisms could arise from nonliving things. This idea was known as **spontaneous generation** or **abiogenesis.** In many cases, this theory arose from misinterpreted observations. Since people often observed maggots (fly larvae) crawling on meat, it was easy for them to believe that meat produced the maggots. Some people believed that snakes came from logs, or that mud produced frogs, or even that old clothes could produce mice, because they observed these animals emerge from those environments. Even Aristotle believed in spontaneous generation. It wasn't until the 1660s that anyone challenged the idea of spontaneous generation and attempted to disprove this notion in a scientific way.

SCIENTISTS AT WORK

Francesco Redi, an Italian physician, tried to learn why maggots appeared on rotting meat. He did not believe that the maggots actually came from the meat. Redi observed the maggots change into small oval structures (pupae) and later into adult flies. From this observation, Redi reasoned that flies had laid their eggs on the meat, and that the eggs had hatched and developed into maggots. He thought that the new generation of flies had to be the offspring of a previous generation of flies.

To test this idea, Redi put meat in both uncovered and covered containers. The meat in the open containers became covered with maggots, but the meat in the closed containers did not. This seemed, to Redi, to disprove the idea of spontaneous generation. However, the supporters of spontaneous generation argued that the maggots did not appear in the closed containers because no air had entered the jars, and everyone knew that air was needed for spontaneous generation to occur. Because of this criticism, Redi redesigned the experiment. This time, he used netting to cover the containers. The netting allowed air in, but the holes in the netting were too small to allow flies to enter and land on the meat. Again, as Redi expected, no maggots appeared on the meat in the containers that were covered with netting. However, the meat in the uncovered containers was once again swarming with maggots.

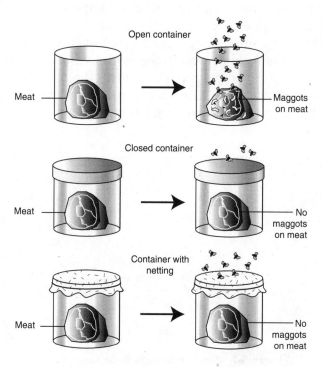

Figure 5-1. The experiment of Francisco Redi.

A few years after Redi's experiments, Anton van Leeuwenhoek, the Dutch lens maker, first saw and described microorganisms. Leeuwenhoek's observations seemed to provide evidence that supported the idea of spontaneous generation, because the tiny microorganisms seemed to appear out of nowhere.

Some scientists continued to perform experiments to try to support the theory of spontaneous generation. In the 1700s, **John Needham,** an English scientist, boiled a batch of broth. He thought that boiling would kill any living organisms in the broth. Needham let the broth stand in flasks sealed with corks. After a few days the broth had become cloudy. An examination showed that the broth was teeming with microorganisms. Needham was quick to declare that the microorganisms in the broth had formed by spontaneous generation.

However, **Lazzaro Spallanzani,** an Italian biologist, disagreed with Needham's conclusion. He felt that Needham did not boil the broth long enough to kill the microorganisms in it, and that the corks had allowed air that carried microorganisms to enter the broth and contaminate it. Spallanzani conducted a series of experiments. He boiled the broth longer than Needham had, and then he melted the glass tops of the flasks to seal them. No new organisms appeared in the broth. Of course, the people who believed in spontaneous generation thought that Spallanzani had prevented air—the "active principle" needed to form microorganisms—from reaching the broth.

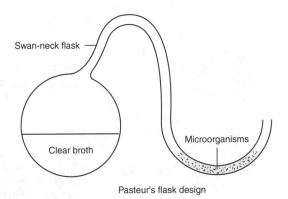

Swan-neck flask

Clear broth

Microorganisms

Pasteur's flask design

Figure 5-2. Pasteur designed the special shape of this flask.

In the 1800s, the final evidence that disproved spontaneous generation was provided by **Louis Pasteur,** the French chemist. Pasteur constructed a flask known as a **swan-neck flask.** The neck of the flask was designed with several curves—like the graceful, curving neck of a swan. Microorganisms in the air would settle into one of the bends in the glass neck and thus not reach the broth. However, air could still pass freely into and out of the flask. Since microorganisms were trapped in the neck of the flask, the broth remained clear. Pasteur's work provided conclusive proof that spontaneous generation, or abiogenesis, did not occur.

You can see in this brief history of an idea how the science of biology changes over time. Today, people believe in **biogenesis,** the idea that living things can come only from other living things. And this belief is an important part of the cell theory: **Living cells come only from existing cells.**

QUESTIONS

Multiple Choice

1. The belief that living things come from nonliving things was known as   *a.* biogenesis   *b.* metabolism
   c. spontaneous generation   *d.* the active principle.

2. The swan-neck flask was designed by   *a.* Pasteur
   b. Redi   *c.* Spallanzani   *d.* Leeuwenhoek.

3. Which of the following tried to prove the theory of abiogenesis?   *a.* Redi   *b.* Needham   *c.* Pasteur
   d. Spallanzani

4. Which of the following scientists first observed and described microorganisms?   *a.* Leeuwenhoek   *b.* Pasteur
   c. Spallanzani   *d.* Redi

5. Redi's results were doubted because of the lack of
   a. water   *b.* a closed container   *c.* space for growth
   d. air.

Matching

6. Redi   *a.* French chemist
7. Needham   *b.* Italian biologist
8. Spallanzani   *c.* Italian physician
9. Pasteur   *d.* Dutch lens maker
10. Leeuwenhoek   *e.* English scientist

Fill In

11. Redi redesigned his experiment by using _____, which allowed air to enter the containers.
12. Most of the scientists involved in the early research to disprove spontaneous generation used liquid _____ to observe the growth of microorganisms.

Portfolio

Louis Pasteur is the subject of many books. His work was instrumental in the development of the modern science of biology. A dramatic film about the life of Pasteur, starring the actor Paul Muni, was also made. See if you can borrow this film from your local library or film-rental store. Write a film review for your school paper.

UNIT THREE
THE CHEMISTRY OF LIFE

Chapter 6
Basic Chemistry

To INVESTIGATE A living organism is to take part in an exercise in wonder that amazes even scientists who have spent many years at this study. A basic understanding of living things requires a knowledge of some of the thousands of chemical activities that occur in each living organism. It is for this reason that it is important to learn some basic concepts of chemistry in a biology course.

MATTER AND CHEMICAL CHANGE

All things, both living and nonliving, are made of matter. **Matter** takes up space and has mass. **Mass** is how much matter is in something. Unlike weight, mass is not dependent on gravity. If you walked on the moon, your weight would be much less than your weight is on Earth. Your mass, the amount of matter that is you, would remain the same on the moon as on Earth. Matter cannot be created or destroyed; however, it can change both chemically and physically. This ability of matter to undergo change is very important to living things.

A **physical change** in matter occurs when there is a change in size, color, or state. For example, you can grind rocks into sand. When rocks are reduced to sand, the basic chemical makeup of the rocks and the sand is the same. Another example of a physical change can be found in the different states of water: as a liquid, as a solid, and as a gas. During evaporation, water can change physically from a liquid to a gas, but it is still water.

A **chemical change** occurs when matter changes from one substance to another. The breakdown of starch into sugars in the body is a good example of a chemical change.

ATOMS: TINY BITS OF MATTER

Atoms are the basic units of matter—the building blocks of all substances. The idea that atoms exist goes back as far as the early Greeks, when **Democritus** talked about matter being made of invisible units. In the early 1800s, **John Dalton,** an English scientist, formulated his atomic theory to explain the "behavior" of matter.

Atoms are so small that they can be seen only with an electron microscope, and even then only the hint of their outline is visible. However, as small as atoms are, they can be broken down into still smaller **subatomic particles.**

There are three main subatomic particles: the proton, the neutron, and the electron. The **proton** is found in the **nucleus,**

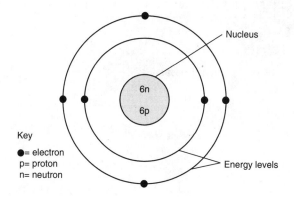

Figure 6-1. The structure of the carbon atom.

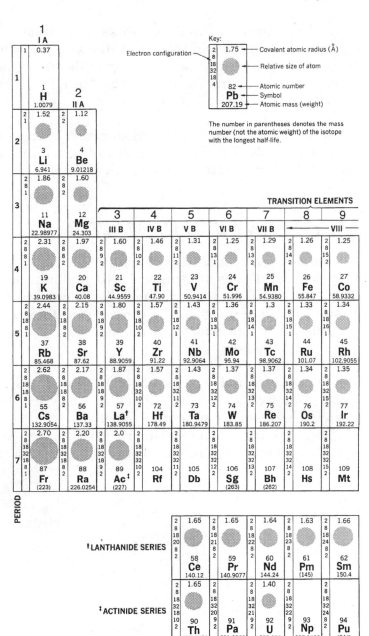

Figure 6-2. The Periodic Table.

	III A	IV A	V A	VI A	VII A	0
						2 / 0.93 — **He** (2) 4.00260
2,3 / 0.88 — 5 **B** 10.81	2,4 / 0.77 — 6 **C** 12.011	2,5 / 0.70 — 7 **N** 14.0067	2,6 / 0.66 — 8 **O** 15.9994	2,7 / 0.64 — 9 **F** 18.998403	2,8 / 1.12 — 10 **Ne** 20.179
2,8,3 / 1.43 — 13 **Al** 26.98154	2,8,4 / 1.17 — 14 **Si** 28.0855	2,8,5 / 1.10 — 15 **P** 30.97376	2,8,6 / 1.04 — 16 **S** 32.06	2,8,7 / 0.99 — 17 **Cl** 35.453	2,8,8 / 1.54 — 18 **Ar** 39.948

I B	II B	III A	IV A	V A	VI A	VII A	0
2,8,16,2 / 1.24 — 28 **Ni** 58.71	2,8,18,1 / 1.28 — 29 **Cu** 63.546	2,8,18,2 / 1.33 — 30 **Zn** 65.38	2,8,18,3 / 1.22 — 31 **Ga** 69.735	2,8,18,4 / 1.22 — 32 **Ge** 72.59	2,8,18,5 / 1.21 — 33 **As** 74.9216	2,8,18,6 / 1.17 — 34 **Se** 78.96	2,8,18,7 / 1.14 — 35 **Br** 79.904 / 2,8,18,8 / 1.69 — 36 **Kr** 83.80

Transition / post-transition continuation:

	I B	II B	III A	IV A	V A	VI A	VII A	0
2,8,18,18 / 1.38 — 46 **Pd** 106.4	2,8,18,18,1 / 1.44 — 47 **Ag** 107.868	2,8,18,18,2 / 1.49 — 48 **Cd** 112.41	2,8,18,18,3 / 1.62 — 49 **In** 114.82	2,8,18,18,4 / 1.40 — 50 **Sn** 118.69	2,8,18,18,5 / 1.41 — 51 **Sb** 121.75	2,8,18,18,6 / 1.37 — 52 **Te** 127.60	2,8,18,18,7 / 1.33 — 53 **I** 126.9045	2,8,18,18,8 / 1.90 — 54 **Xe** 131.30
2,8,18,32,17,1 / 1.38 — 78 **Pt** 195.09	2,8,18,32,18,1 / 1.44 — 79 **Au** 196.9665	2,8,18,32,18,2 / 1.55 — 80 **Hg** 200.59	2,8,18,32,18,3 / 1.71 — 81 **Tl** 204.37	2,8,18,32,18,4 / 1.75 — 82 **Pb** 207.2	2,8,18,32,18,5 / 1.46 — 83 **Bi** 208.9804	2,8,18,32,18,6 / 1.4 — 84 **Po** (209)	2,8,18,32,18,7 / 1.40 — 85 **At** (210)	2,8,18,32,18,8 / 2.2 — 86 **Rn** (222)

Lanthanide series:

2,8,18,25,8,2 / 1.65 — 63 **Eu** 151.96	2,8,18,25,9,2 / 1.61 — 64 **Gd** 157.25	2,8,18,27,8,2 / 1.59 — 65 **Tb** 158.9254	2,8,18,28,8,2 / 1.59 — 66 **Dy** 162.50	2,8,18,29,8,2 / 1.58 — 67 **Ho** 164.9304	2,8,18,30,8,2 / 1.57 — 68 **Er** 167.26	2,8,18,31,8,2 / 1.56 — 69 **Tm** 168.9342	2,8,18,32,8,2 / 1.70 — 70 **Yb** 173.04	2,8,18,32,9,2 / 1.56 — 71 **Lu** 174.967

Actinide series:

2,8,18,32,25,8,2 — 95 **Am** (243)	2,8,18,32,25,9,2 — 96 **Cm** (247)	2,8,18,32,26,9,2 — 97 **Bk** (247)	2,8,18,32,28,8,2 — 98 **Cf** (251)	2,8,18,32,29,8,2 — 99 **Es** (254)	2,8,18,32,30,8,2 — 100 **Fm** (257)	2,8,18,32,31,8,2 — 101 **Md** (258)	2,8,18,32,32,8,2 — 102 **No** (259)	2,8,18,32,32,9,2 — 103 **Lr** (260)

or core of the atom. The proton has a positive charge and is assigned a mass of 1 atomic mass unit, or 1 AMU. The **atomic number** of an atom is equal to the number of protons in that atom. The **neutron,** which has no charge, is also located in the nucleus and assigned a mass of 1 AMU. When we add the mass of the protons to the mass of the neutrons in an atom, we find what is called the **atomic mass.**

The third main subatomic particle, the **electron,** is not located in the nucleus. It is in an area around the nucleus. Electrons carry a negative charge and have a very slight mass (1/1836 AMU). Electrons were discovered by **Joseph Thomson,** an English scientist, in 1897. Electrons are located in a space around the nucleus known as the **electron cloud.** Within this cloud are energy levels that contain the electrons. The electrons are able to move anywhere in the cloud, and since they move so quickly, it is impossible to predict where an individual electron will be found at any particular instant. The number of protons and the number of electrons in an atom are equal. Since there are an equal number of positive charges and negative charges, the atom is neutral.

ELEMENTS AND ISOTOPES

An **element** contains only one type of atom. Ninety-two elements occur naturally, but other elements have been produced in laboratories. In fact, new elements are produced every year. However, some of these laboratory elements exist for only a fraction of a second. Six elements—oxygen, carbon, hydrogen, nitrogen, phosphorus, and sulfur—make up approximately 99 percent of the human body.

A kind of symbolic shorthand is used by scientists when talking or writing about elements. For example, carbon is represented by the letter C, oxygen by the letter O. Sometimes, more than one letter is used as the symbol for an element. The first letter is always capitalized, the second never is. When elements are arranged by their atomic masses, they fall into groups that show similarities in their properties. This arrangement is called the **periodic table.** A knowledge of the periodic table is essential in the study of chemistry. You can predict how the atoms of an element will behave by knowing the element's position in the periodic table.

Atoms of an element can exist in different forms. Atoms of the same element may have the same number of protons and electrons but different numbers of neutrons. These atoms are called **isotopes.** Carbon-12 and Carbon-14 are isotopes. Carbon-12 has six protons and six neutrons, and Carbon-14 has six protons and eight neutrons. As a result, they each have a different atomic mass. Certain isotopes are used to date human artifacts; other isotopes have important uses in medicine.

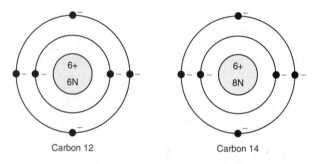

Carbon 12   Carbon 14

Figure 6-3. Isotopes of carbon.

COMPOUNDS AND BONDS

A **compound** is formed when two or more atoms of different elements combine in a definite proportion. An example of a compound is NaCl (sodium chloride). This is the compound known as table salt, or sea salt. This common compound forms when a single sodium (Na) atom combines with a single chlorine (Cl) atom.

To form compounds, atoms must be held together. The holding together of atoms is accomplished by **chemical bonds.**

One type of chemical bond is an **ionic bond.** An ionic bond is formed when one atom gives up one or more electrons to another atom. When this happens, both atoms become electrically stable. However, as a result of losing or gaining electrons, each atom now has an electrical charge. It is the opposite electrical charges that cause the atoms to bond together.

Covalent bonds are another type of chemical bond. In covalent bonds, two atoms share two or more electrons. This sharing of electrons holds the atoms together. A **molecule** is formed when two or more atoms are held together by covalent bonds. Examples of molecules are carbon dioxide (CO_2) and water (H_2O).

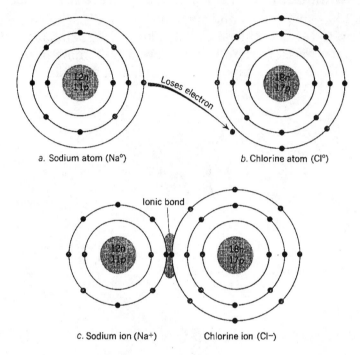

a. Sodium atom (Na°)   *b.* Chlorine atom (Cl°)

c. Sodium ion (Na+)   Chlorine ion (Cl−)

Figure 6-4. An ionic bond formed by sodium and chlorine.

We can use molecular formulas or structural formulas to show molecules. A **molecular formula** shows the kinds and numbers of atoms in the molecule. The molecular formula H_2O represents water. A molecular formula is sometimes called a

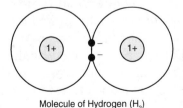

Molecule of Hydrogen (H_2)

Figure 6-5. A covalent bond formed by two atoms of hydrogen.

chemical formula. **Structural formulas** indicate how the atoms are arranged. The structural formula for water is H-O-H.

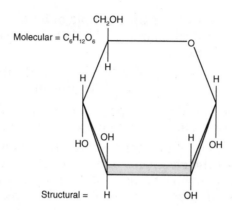

Figure 6-6. The structural and molecular formulas for glucose.

A **chemical reaction** occurs when chemical bonds are made or broken. The following equation shows an example of a chemical reaction:

$$2H_2 + O_2 \longrightarrow 2H_2O$$

The starting materials, hydrogen and oxygen ($2H_2 + O_2$), are called the **reactants.** The final material, water ($2H_2O$), is the **product.** The arrow indicates the direction of the reaction. Thousands of chemical reactions occur every second in living things.

When some substances are mixed together they do not combine chemically. Instead, these substances form **mixtures.** Mixtures may be homogeneous or heterogeneous. In a **homogeneous mixture,** the substances are distributed in such a way that each part of the mixture looks the same. A **solution** is a homogeneous mixture. Salt water is an example of a solution. In this case, salt particles are pulled apart by the water molecules. In a **heterogeneous mixture,** areas of the mixture look different. **Suspensions** are heterogeneous mixtures. Soil mixed with water is a suspension. The mixture eventually separates, and two layers, water and soil, form. The heavier soil layer would collect beneath the lighter layer of water.

QUESTIONS

Multiple Choice

1. The freezing of water is an example of a   *a.* chemical change   *b.* physical change   *c.*   chemical reaction   *d.* creation of matter.

2. The atomic theory was developed by   *a.* Dalton   *b.* Democritus   *c.* Einstein   *d.* Thomson.

3. Matter made of only one type of atom is called   *a.* an isotope   *b.* a compound   *c.* an element   *d.* a mixture.

4. The atomic mass of an atom is the   *a.* mass of the proton + the mass of the electron   *b.* mass of the electron + the mass of the neutron   *c.* mass of the proton + the mass of the nucleus   *d.* mass of the proton + the mass of the neutron.

5. Electrons are located in an atom's   *a.* nucleus   *b.* core   *c.* cloud   *d.* area of greatest mass.

Matching

6. matter
7. element
8. compound
9. isotope
10. mixture

a. matter that contains only one type of atom

b. atoms with the same number of protons but different numbers of neutrons

c. anything that takes up space and has mass

d. two or more atoms of different elements that are combined chemically

e. substances put together that do not combine chemically

Fill In

11. _____ are the basic units of matter.

12. Protons, neutrons, and electrons are all _____ particles.

13. When two or more atoms of different elements combine, a _____ is formed.

14. The starting materials in a chemical reaction are called the _____ .

15. A solution is a type of _____ mixture.

Portfolio

1. Make models to show the atomic structure of a few common compounds. You can use pipe cleaners and plastic foam balls or pipe cleaners and gumdrops.

2. Interview a physician to find out ways that isotopes are used in hospitals to diagnose and treat certain diseases. You may want to present your findings to your class.

Chapter 7

Organic Compounds

ON A BASIC level, all living things are made of elements, as are all nonliving things. However, the thin line that separates living and nonliving things is, in many ways, wider than the Grand Canyon. Scientists have determined the chemical makeup of many of the substances found in living organisms, but they cannot put those chemicals together in a way that would make them alive.

Most of the compounds that make up living things are **organic compounds.** Organic compounds contain the element **carbon.** Because of the structure of its atoms, carbon has the ability to form four covalent bonds. This allows carbon to form some compounds that are shaped like chains and others that are shaped like rings. These structures can be simple, consisting of only a few atoms, or very complex, consisting of many atoms. Luckily for living things, most carbon compounds are very stable.

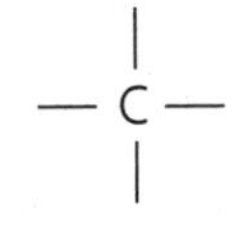

Figure 7-1.
A carbon atom.

Even though carbon is the main element, organic compounds in living things also contain hydrogen, oxygen, nitrogen, sulfur, and phosphorus. In living things, these organic compounds are usually very large, often consisting of hundreds or thousands of atoms. The study of organic compounds is called **organic chemistry.** A few carbon compounds, such as CO_2, and all compounds without carbon are called **inorganic compounds.**

MONOMERS AND POLYMERS

Some organic molecules are small and are called **monomers**. The most simple organic compound is the gas methane, CH_4. Methane is called natural gas, or swamp gas. A molecule of methane consists of one carbon and four hydrogen atoms. Monomers may be linked together to form **polymers**. Polymers can be extremely long molecules that contain many monomers. Polymers are also called **macromolecules**. One way monomers link together is by a **condensation reaction**. In a condensation reaction, monomers lose hydrogen and oxygen atoms, which form water. The loss of these atoms opens bonding sites. Monomers can attach to each other at these sites. Because condensation reactions cause a loss of water from the monomers when they link together, these reactions are also called **dehydration synthesis reactions.**

Figure 7-2. Methane, a simple organic molecule, is also known as natural gas.

A condensation reaction can also be reversed. When water is added at a polymer's bonding sites, the monomers separate. This process is called **hydrolysis.** Hydrolysis breaks polymers down into monomers. Condensation and hydrolysis are two important reactions in the formation and breakdown of organic compounds in living things.

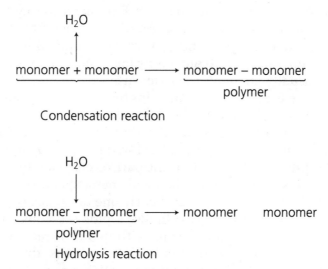

Figure 7-3. Condensation and hydrolysis.

TYPES OF ORGANIC COMPOUNDS

There are four types of organic compounds found in living things: proteins, lipids, nucleic acids, and carbohydrates. **Proteins** are polymers that are made of monomers called **amino acids.** Proteins are an important component of muscles and many other structures in the body. Proteins are important in the body's building processes. They contain carbon, hydrogen, oxygen, and nitrogen. Although there are only 20 different amino acids, there are thousands of different proteins. This tremendous variety in the kinds of proteins is possible because the 20 amino acids can link together in so many different ways.

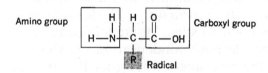

Figure 7-4. A general formula for the structure of an amino acid.

Enzymes are a very special group of proteins that are made in cells. Enzymes are organic **catalysts,** chemicals that alter the speed of a chemical reaction but remain unchanged themselves. Enzymes are very specific in their activity; each enzyme causes only one or a few reactions to occur. For enzymes to work, they must bind to a substrate. A **substrate** is the reactant the enzyme affects. When an enzyme and substrate combine, the product is called an **enzyme-substrate complex.** The names of specific enzymes usually end with the letters *ase*, such as maltase or lactase. Maltase is an enzyme that helps break down the sugar maltose; lactase breaks down milk sugar, or lactose.

Lipids are organic molecules that have three major functions in organisms. Energy can be stored in the body in the form of lipids. Lipids are an important part of cell membranes; and lipids can also be used as chemical messengers. Steroids are common lipids. Cholesterol and cortisone are types of steroids. Fats, oils, and waxes are also lipids. Lipids are constructed by combining one glycerol molecule with two or three fatty acids. Glycerol is an organic molecule that contains three carbon atoms, each attached to a hydroxyl (-OH) group. A fatty acid consists of a long chain of carbon atoms attached to a carboxyl

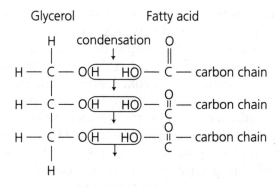

Figure 7-5. A lipid.

(COOH) group. One property of lipids is that they are insoluble in water.

Another group of complex organic molecules are the **nucleic acids.** DNA (deoxyribonucleic acid) and RNA (ribonucleic acid) are nucleic acids. DNA, a double-helix-shaped molecule, is important because it stores the genetic information of the cell. RNA helps direct the building of proteins by determining the order in which amino acids are linked. Both DNA and RNA are polymers made of thousands of monomers called **nucleotides.** Nucleotides are composed of a phosphate, a sugar, and a base.

Carbohydrates are made of carbon, hydrogen, and oxygen. In all carbohydrates, the hydrogen and oxygen atoms are found in a 2:1 ratio. This is the same ratio of hydrogen to oxygen atoms

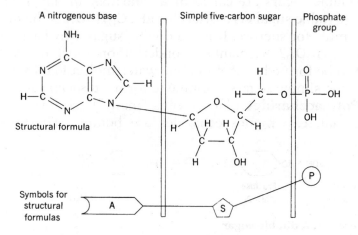

Figure 7-6. The structure of a nucleotide.

found in water molecules. Carbohydrates are a source of energy for living things. Sugars and starches are carbohydrates.

Carbohydrates may be simple or complex. **Monosaccharides** (*mono-* means "one," *saccharide* means "sugar") are simple sugars. They cannot be broken down into smaller sugars. Glucose is a monosaccharide. Its molecular formula is $C_6H_{12}O_6$. Fructose and galactose are also monosaccharides. They have the same molecular formula as glucose, $C_6H_{12}O_6$, but they have different structural formulas. When a substance has the same molecular formula but a different structural formula, it is called an **isomer.** Fructose and galactose are isomers of glucose.

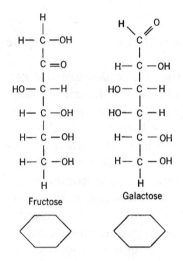

Figure 7-7. Two isomers of glucose.

Double sugars are called **disaccharides** (*di* means "two"). Sucrose (table sugar) is a disaccharide. $C_{12}H_{22}O_{11}$ is the molecular formula for sucrose. If it is a double sugar, why isn't the formula $C_{12}H_{24}O_{12}$? Remember, condensation must occur for a bond to be formed; therefore, one water molecule (H_2O) is lost when two simple sugars combine to form a disaccharide.

Polysaccharides (*poly-* means "many") are formed when more than two monosaccharides are bonded. They may be

HO⟨⟩OH + HO⟨⟩OH —enzyme→ HO⟨—O—⟩OH + H_2O

Glucose   Fructose   Sucrose
$C_6H_{12}O_6$ $C_6H_{12}O_6$ $C_{12}H_{22}O_{11}$

Figure 7-8. A double sugar.

made of hundreds of monosaccharides or even thousands of monosaccharides. Starch (stored sugar in plants) and cellulose (a structural molecule found in the cell walls of plants) are polysaccharides. They are composed of many glucose molecules. Cellulose is the most abundant organic compound on Earth. Glycogen is another common polysaccharide. In the human body, glycogen is stored in the liver and is broken down into glucose molecules when the body's sugar level is too low. Another important polysaccharide is chitin. It is used by arthropods, like insects, to make their exoskeletons.

QUESTIONS

Multiple Choice

1. The study of organic compounds is called *a.* inorganic chemistry *b.* organic chemistry *c.* physical chemistry *d.* applied chemistry.
2. Polymers are made of units called *a.* peptides *b.* macromolecules *c.* isomers *d.* monomers.
3. Which of the following reactions must occur in order to break down polymers into monomers? *a.* hydrolysis *b.* isomerism *c.* dehydration synthesis *d.* condensation.
4. Proteins are polymers made of monomers called *a.* saccharides *b.* amino acids *c.* lipids *d.* peptides.
5. Enzymes are a type of *a.* carbohydrate *b.* protein *c.* lipid *d.* nucleic acid.

Matching

6. sucrose
7. glucose
8. amino acid
9. starch
10. steroid

a. a monosaccharide
b. a disaccharide
c. a polysaccharide in plants
d. a lipid in animals
e. a monomer in proteins

Fill In

11. Enzymes are organic _____, which help speed up chemical reactions.
12. Lipids are a source of _____ for the cell.
13. Organic compounds always contain the element _____.
14. Nucleic acids are made up of many monomers called _____.
15. The molecular formula for glucose is _____.

Portfolio

A low-fat diet is considered important for good health. But a very low-fat diet may be dangerous. Can you explain why this statement is true? You may have to go to the library to find out more about low-fat diets. Report your findings to the class.

UNIT FOUR
CELLS: THE UNITS OF LIFE

Chapter 8

Cell Structure and Function

Humans are multicellular organisms, made up of trillions of cells. In fact, there are about 30 billion cells in your brain alone. Some of these cells help you think about life. The cells in your brain also help you read and understand the material in this book. Other brain cells help coordinate your movements; still others receive information about the environment gathered by your ears, eyes, and the touch receptors in your skin.

Like humans, many organisms consist of millions of cells. Some organisms consist of just a few cells, while others consist of only a single cell. Cells are the basic unit of organization for all organisms. But what is a cell? And how does a cell function?

Scientists have found that all organisms consist of cells. This generalization is known as the cell theory, and was formulated by the German biologists **Schleiden** and **Schwann.** The study of cells is now known as **cytology.**

CELL TYPES

Cells are classified into two different groups. The simplest cells are **prokaryotic,** which means "before the nucleus." Prokaryotic cells lack a membrane-bound nucleus. Bacterial cells are examples of prokaryotic cells.

Other cells are **eukaryotic,** meaning "true nucleus." These cells have a membrane-bound nucleus. The membrane that encloses the nucleus is called the **nuclear membrane.** Most cells contain only one nucleus. The **nucleus** acts as the control center for the cell. It was first identified by the Scottish botanist **Robert Brown.**

The nucleus contains **chromosomes,** the structures that carry the cell's hereditary instructions. Chromosomes are made of DNA and proteins. The **nucleolus,** which plays a part in protein synthesis, is also found in most nuclei.

CELL ORGANELLES

To survive, each cell must be able to carry out all functions necessary for life. It must be able to absorb, transport, digest, synthesize, secrete, respire, excrete, reproduce, and respond. Cells contain a variety of membrane-bound "little organs" called **organelles.**

All cells are enclosed by a **cell membrane.** The cell membrane is composed largely of proteins and lipids. It separates the contents of the cell from the extracellular fluid (ECF). The cell membrane forms the cell's boundary. The function of the cell membrane is to control the movement of substances into and out of the cell. The cell membrane provides a **homeostatic,** or balanced, environment inside the cell. The cell membrane is **semipermeable,** or selectively permeable. It allows small molecules to pass through while blocking the passage of large molecules.

A fluidlike substance called **cytoplasm** is found in prokaryotic cells. Cytoplasm is also found between the nuclear membrane and cell membrane of eukaryotic cells. Cytoplasm is a thick, clear material with various organelles suspended in it.

Ribosomes are small spherical organelles that serve as the site of protein synthesis. Ribosomes can be found in both prokaryotic and eukaryotic cells. They are among the most numerous of cellular organelles. In eukaryotic cells, some ribosomes are attached to the **endoplasmic reticulum** (ER). The ER serves as a site for chemical reactions. It also functions as a

Table 8–1.   Cell Organelles and Their Functions

Organelle	Functions
Cell wall	Protects and supports cell
Centrosome and centrioles	Aid in cell division
Chloroplasts	Carry on photosynthesis
Chromosomes	Contain the genetic material (DNA) of the cell
Cilia and flagella	Enable locomotion, movement of fluids
Cytoplasm	Stores chemicals; carries on anaerobic respiration
Cytoskeleton	Structural network enables movement of cell and organelles within the cell
Endoplasmic reticulum	Transports substances within cell; protein synthesis
Golgi apparatus	Stores secretions; modifies and refines proteins
Lysosome	Digests materials in cell
Mitochondria	Release energy
Nuclear membrane	Regulates passage of substances into and out of nucleus
Nucleolus	Stores RNA
Nucleus	Controls all cell activities, including reproduction
Cell membrane	Regulates passage of water and dissolved substances into and out of cell
Ribosomes	Synthesize proteins
Vacuoles	Store water and dissolved substances

major transport system for proteins. The ER connects the nuclear membrane to the cell membrane. If ribosomes are attached to the ER, it is called rough ER. Smooth ER does not have any ribosomes attached to it.

ORGANELLES IN EUKARYOTIC CELLS

Eukaryotic cells contain a number of organelles that are not found in prokaryotic cells. For example, **lysosomes** are organelles in animal cells that are responsible for digestion and waste disposal. Lysosomes contain enzymes that break down large molecules and destroy worn-out cell structures. Lysosomes also cause injured cells to self-destruct. The **Golgi apparatus** is another type of organelle. It is made of several membrane-bound, flattened sacs. These structures package and secrete cell products. **Mitochondria** are known as the "powerhouses of the cell." They contain DNA, RNA, and enzymes, and they are enclosed by two membranes. Cellular respiration, the process that releases the chemical energy stored in food, takes place in the mitochondria. Thousands of mitochondria are present in cells that require a lot of energy to function, such as muscle cells.

Vacuoles are organelles that are usually used for the storage of water, proteins, salts, and carbohydrates. Vacuoles may also digest food, contain pigments, and dispose of wastes. In plants, vacuoles that are filled with water help support tissues. **Centrioles** most often lie just outside the nucleus and are important in cell reproduction. A **cytoskeleton** that consists of a variety of filaments and fibers supports cell structure and helps the cell move.

PLANT AND ANIMAL CELLS

A distinction can be made between plant and animal cells based on the organelles and structures they contain. Plant cells have a large central vacuole called a **tonoplast.** This large vacuole produces internal pressure against the cell wall. Plant cells also contain the green pigment chlorophyll, packaged in an

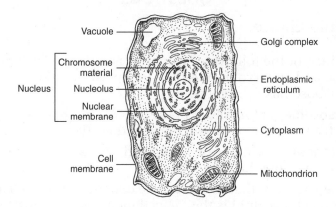

Figure 8-1. Structures found in a typical animal cell.

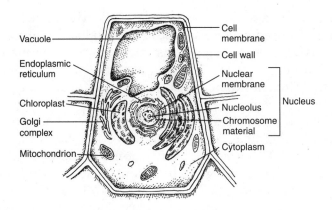

Figure 8-2. Structures found in a typical plant cell.

organelle called a **chloroplast.** Chlorophyll traps the energy in light to be used in the all-important function of photosynthesis. The rigidity of plant cells is due to the nonliving **cell wall** found bordering the outside of the less rigid cell membrane. The cell wall provides protection and support and is made up of two or more layers. The cell wall is composed mainly of cellulose (a carbohydrate) and is more rigid than the cell membrane. Cell walls are also found in algae and in some kinds of bacteria. Animal cells do not have a central vacuole, chloroplasts, or cell walls.

QUESTIONS

Multiple Choice

1. Which of the following structures is found in both prokaryotic and eukaryotic cells?   *a.* tonoplast   *b.* lysosomes   *c.* ribosomes   *d.* centrioles.

2. Lysosomes are involved in the process of   *a.* respiration   *b.* transport   *c.* reproduction   *d.* digestion.

3. A tonoplast is a type of   *a.* plastid   *b.* vacuole   *c.* membrane   *d.* plant cell.

4. The structure that separates the contents of the cell and the extracellular fluid is the   *a.* cell membrane   *b.* ER   *c.* nuclear membrane   *d.* mitochondria.

5. Muscle cells need a tremendous amount of energy. Therefore, they contain a large number of   *a.* ribosomes
   b. Golgi apparatus   *c.* vacuoles   *d.* mitochondria.

6. The process of photosynthesis occurs in the   *a.* mitochondria   *b.* chloroplast   *c.* vacuole   *d.* ribosome.

7. After proteins are synthesized in the ribosomes, which organelle transports the proteins throughout the cell?
   a. lysosomes   *b.* vacuoles   *c.* endoplasmic reticulum
   d. cell membrane.

Organizing Data

8. Complete the table.

Shape	Organelle	Function
_____	_____	transport
_____	ribosome	_____
_____	_____	digestion
_____	chloroplast	_____
_____	_____	makes energy
_____	Golgi apparatus	_____
_____	vacuoles	_____

9. Place a check in the appropriate column.

	FOUND IN	
Structure	Prokaryotes	Eukaryotes
cytoplasm	_____	_____
nucleus	_____	_____
mitochondria	_____	_____
ribosome	_____	_____
Golgi apparatus	_____	_____
endoplasmic reticulum	_____	_____
lysosome	_____	_____
vacuole	_____	_____

Fill In

10. The large vacuole in a plant cell is called a _____.
11. The site of protein synthesis is the _____.
12. Cells that lack a nucleus are called _____ cells.
13. Ribosomes may be attached to another organelle called the _____.
14. Plant cells have a _____ _____ outside the cell membrane.
15. Chromosomes are found in the _____ nucleus of a eukaryotic cell.

Free Response

16. What are some ways that animal and plant cells are different?
17. Why is it important for a cell to have a semipermeable membrane?
18. What are some differences between prokaryotic and eukaryotic cells?

Portfolio

Make a drawing that shows five important structures found in an animal cell. Briefly tell how the cell you drew would be different if it lacked each of these five organelles.

Chapter 9

Homeostasis and Transport in a Cell

$\mathbf{Y}$OU KNOW HOW important balance is when you watch a gymnast perform. A slight move in the wrong direction could spell disaster in a much-practiced routine. In many ways, organisms must maintain a delicate internal balance in order to survive. A biochemical move in the wrong direction could mean death.

In order for cells to live, they must be in balance with the environment around them. This balance is known as **homeostasis.** Since the cell membrane regulates what goes in and out of a cell, it plays an important role in maintaining this balance. The chemical laws that describe the movement of molecules help us understand how homeostasis is reached.

THE DIFFUSION OF MOLECULES

Molecules are in constant motion. When there are lots of molecules in an area, it is known as a **high concentration** of molecules. When there are only a few, it is called a **low concentration** of molecules. Molecules tend to move from areas of high concentration to areas of low concentration. This movement is termed **diffusion.** Diffusion occurs in gases, liquids, and solids.

A good example of the diffusion of gases occurs when you open a bottle of perfume. There is a very high concentration of

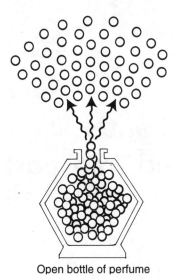

Open bottle of perfume

Figure 9-1. Molecules of perfume in an open bottle move into the air.

perfume molecules in the bottle and a low concentration outside of the bottle. There is a high concentration of air molecules outside of the bottle and a low concentration inside the bottle. When the top of the bottle is removed, perfume molecules move out of the bottle, and air molecules move into the bottle. Each type of molecule moves from an area of high concentration to an area of low concentration of its molecules.

If you add a few drops of food coloring to a glass of water, you can observe diffusion occurring in a liquid. It is easy to see the movement of food coloring in the water. The wispy trails of color are actually molecules of food coloring diffusing through the molecules of water. Of course, you cannot see the molecules themselves moving, only their colorful trail. Eventually the molecules of food coloring will be distributed evenly throughout the water. And if you have added enough food coloring, in time the water will be evenly tinted a paler version of that color.

DIFFUSION THROUGH A MEMBRANE

Diffusion can also occur through a solid if it contains open spaces. A solid with open spaces is **permeable.** A cell membrane allows some things to pass through while stopping others. This type of membrane is called **semipermeable** or **selectively permeable.** For example, water molecules are small enough to pass through cell membranes, but large molecules such as starch cannot pass through cell membranes.

Carrier molecules in a cell membrane can help in the process of diffusion. By combining with a molecule to be transported, a carrier molecule can help move the molecule across a cell membrane. This process is called **facilitated diffusion.** No extra energy is needed, and the normal rules of diffusion are followed (molecules move from an area of high concentration to an area of low concentration).

The rate, or speed, of diffusion can be affected by several factors. An increase in temperature causes the molecules to move faster; as a result, the rate of diffusion increases. An increase in pressure also causes the rate of diffusion to increase.

OSMOSIS

Water molecules continually enter and leave cells by **osmosis,** a kind of diffusion. The diffusion of water molecules through a semipermeable membrane occurs from an area of high concentration of water molecules to an area of low concentration of water molecules.

The environment that surrounds a cell determines the direction in which water molecules move. When there are more water molecules outside a cell than inside a cell, water molecules move into the cell.

A solution that contains more water molecules than are located in the cell is called a **hypotonic solution.** A cell in this type of solution begins to swell as water moves into it. The water molecules move into the cell (the area of low concentration) from the hypotonic solution (the area of high concentration of water molecules).

If a cell is placed in a **hypertonic solution,** a solution that has fewer water molecules outside the cell than are inside the cell, water molecules move out of the cell. As the cell loses water molecules, it begins to shrivel up.

There is a point in the movement of molecules when the rate of molecules leaving the cell becomes equal to the rate of molecules entering the cell. This is called the point of equilibrium. In an **isotonic solution,** the number of water molecules outside the cell is equal to the number of molecules inside the

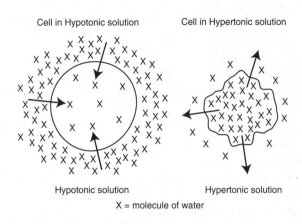

Figure 9-2. Hypotonic and hypertonic solutions.

cell. Therefore there is no movement of molecules into or out of the cell.

In plant cells, water molecules press against the inside of the cell wall. This pressure is called **turgor pressure** and helps to support a plant's tissues. A cell that has a strong turgor pressure is termed **turgid.** When a plant cell loses water molecules, pressure is lost. This loss of water is called **plasmolysis.** The loss of water by many cells in the plant causes wilting.

Passive and Active Transport

Both diffusion and osmosis are types of **passive transport.** This means that they do not use energy to help with the movement of molecules. Sometimes molecules must be moved against the direction of diffusion in order for the cell to reach homeostasis. This movement against normal diffusion requires energy. Movement that requires energy is called **active transport.**

In other situations, it is important for a cell not to reach homeostasis. For example, the proper functioning of nerve cells depends on active transport to maintain an imbalance in sodium and potassium ions. The sodium-potassium pump is used by nerve cells to maintain the proper levels of sodium and potassium ions in the cells. This pump transports sodium and potassium ions across the cell membrane in such a way that homeostasis is not reached. This allows electrical charges to be

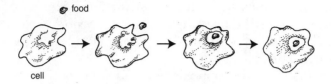

Figure 9-3. Steps in phagocytosis.

created across the membrane that enable an impulse to move along a nerve cell.

Cells are also able to bring in and excrete molecules that are too large to pass through the openings in their cell membrane. **Endocytosis** is the process that allows a cell to bring in materials from its surroundings. There are two types of endocytosis. **Phagocytosis** allows cells to engulf large particles. **Pinocytosis** allows a cell to take in liquid molecules. Pinocytosis is accomplished when the cell membrane folds inward, pulling the

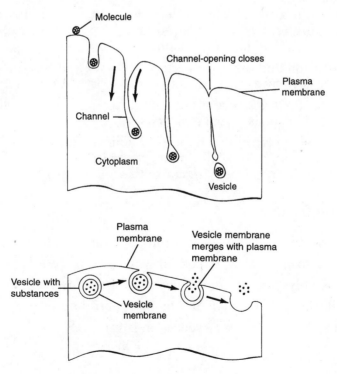

Figure 9-4. Steps in pinocytosis.

dissolved molecules in with it as it forms a depression in the membrane. Later the opening to the depression is closed, trapping the liquid. The bottom of the depression breaks off to form a vacuole in the cell.

Exocytosis is the opposite of endocytosis. **Exocytosis** allows materials to be discharged and secreted from a cell. For example, proteins produced in the ribosomes of cells are transported out of the cells by exocytosis.

QUESTIONS

Multiple Choice

1. In diffusion, molecules move from   *a.* areas of low concentration to areas of high concentration   *b.* areas of high concentration to areas of low concentration   *c.* isotonic areas to hypertonic areas   *d.* passive areas to active areas.

2. The diffusion of water molecules through a semipermeable membrane is called   *a.* equilibrium   *b.* homeostasis   *c.* osmosis   *d.* active transport.

3. Diffusion and osmosis are types of   *a.* equilibrium   *b.* passive transport   *c.* active transport   *d.* plasmolysis.

4. Pinocytosis is a type of   *a.* endocytosis   *b.* exocytosis   *c.* phagocytosis   *d.* plasmolysis.

5. If a cell is placed in a hypotonic solution, water will   *a.* move out of the cell   *b.* not move in either direction   *c.* move into the cell   *d.* cause the cell to shrink.

Matching

6. pinocytosis	*a.* drooping of a plant stem
7. exocytosis	*b.* loss of water from a cell
8. plasmolysis	*c.* lots of water pressure
9. wilting	*d.* taking in of liquid molecules
10. turgid	*e.* secretion from a cell

Fill In

11. The balance between a cell and its external environment is called _____.
12. A membrane that allows some molecules to pass through but stops larger molecules is termed _____.
13. When a cell is placed in a _____ solution, water molecules move out of the cell.
14. The process by which water molecules move through a membrane is called _____.
15. Carrier molecules are involved in the process of _____ diffusion.

Portfolio

Observe a situation on a bus or train or in a public waiting room. Where do people stand or sit when the area is mostly empty? Are people approximately evenly spaced from one another? What happens as the space fills up? What happens when the space is crowded? What happens when the space empties out again? Think of the people as models for molecules. How does this situation remind you of the processes of osmosis and diffusion? You might like to describe your observations or draw pictures to represent your observations.

Chapter 10

Energy and Cellular Processes

Look at people who have just finished running a marathon. You can see the effort of completing this task reflected in their faces. Many collapse at the finish line, too weak to stand. Few marathon runners have enough energy left to complain about their tired condition.

Running a race requires a great deal of energy. But sitting still requires some energy, too. All living things require energy. Each cell needs energy to carry on all the activities that keep it alive and healthy. Cells get most of their energy from carbohydrates such as sugars and starches. The energy can be used immediately or stored for later use. A marathon race is so long and grueling, however, that almost all the energy stored in a runner's body is used up.

ATP: STORED ENERGY

When energy is stored it is usually stored in the energy carrier **ATP** (adenosine triphosphate). ATP is found in all living things and is able to absorb energy and release it when needed by a cell. ATP is made up of one adenosine molecule and three

A – P ~ P ~ P
↑
High-energy bond

Figure 10-1. Energy is stored in high-energy bonds.

$$A - P \sim P \lessgtr P \longrightarrow A - P \sim P + \text{Energy}$$

$$ATP \longrightarrow ADP + \text{Energy}$$

Figure 10-2. When a bond is broken, energy is released.

phosphate groups. The bond holding together the last two phosphates is a high-energy bond. When the bond is broken, a tremendous amount of energy is released for use by the cell. After the bond is broken, the remaining molecule holds only two phosphate groups. It is now called **ADP** (adenosine diphosphate). ADP can absorb energy, use this energy to add another phosphate, and become ATP again. ATP allows the cell to use its energy a little at a time.

CELLULAR RESPIRATION

Where does the cell get the energy to change ADP into ATP? The carbohydrate **glucose** is the cell's main energy source. Glucose contains energy in its chemical bonds. The breakdown of glucose is called **cellular respiration.** Cellular respiration is the process that powers the cell. The reaction for cellular respiration is represented by the following equation:

$$C_6H_{12}O_6 + 6O_2 \longrightarrow 6CO_2 + H_2O + ATP$$

In this process, glucose and oxygen combine to form carbon dioxide, water, and energy (ATP). Since oxygen is involved, the process is called aerobic respiration.

Aerobic respiration occurs in two stages. The first stage is called glycolysis and occurs in the cell's cytoplasm. Glycolysis involves a reaction that splits a molecule of glucose into two molecules of pyruvic acid—a three-carbon compound. Two molecules of ATP are required to get this reaction started. However, four ATP molecules are produced, giving glycolysis a net production of two ATP molecules. Since oxygen is not involved in the first stage of glycolysis, it is termed anaerobic respiration.

The second stage of cellular respiration is called the citric acid cycle or the Krebs cycle. This second stage completes the breakdown of glucose and requires oxygen. The pyruvic acid

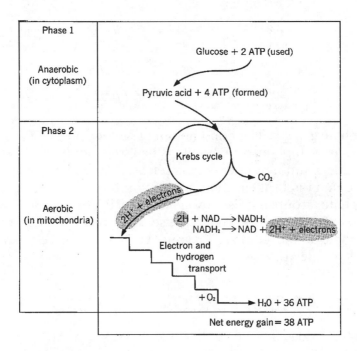

Figure 10-3. Energy release in cells.

made in glycolysis is used in the Krebs cycle. Now the pyruvic acid is changed into CO_2, H_2O, actual ATPs, and potential ATPs. CO_2 is a waste product and is released by the cell. Two actual ATPs are made in the Krebs cycle. The potential ATPs are converted into actual ATPs by a series of electron carriers called the ETC (electron transport chain). Eventually, a net gain of 36 molecules of ATP is produced in the process of cellular respiration. These are the ATP molecules that provide the energy needed to run the cell.

FERMENTATION

In some cases, glucose is broken down in a process that doesn't use oxygen. This process is called **fermentation**. The first stage of fermentation is identical to the first stage of cellular respiration. Glycolysis converts glucose into two pyruvic acids and produces two ATPs. Then, unlike normal cellular respiration, the pyruvic acid produces either lactic acid or alcohol

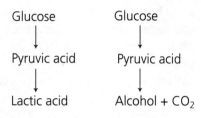

Figure 10-4. Fermentation.

and CO_2. If lactic acid is formed, the process is called **lactic acid fermentation.** If alcohol and CO_2 are produced, it is called **alcoholic fermentation.**

Lactic acid fermentation can take place in humans when not enough oxygen is present in our bodies. A good example of this type of fermentation occurs when muscles are used continuously. After a period of strenuous exercise, such as running a marathon, there may not be enough oxygen left to continue the normal aerobic respiration that provides the energy needed by muscle cells. As a result, fermentation (anaerobic respiration) takes place, causing lactic acid to build up in the muscles. This causes the muscles to tire. At this stage the muscle cells need time to gather the much-needed oxygen. Think of how deeply you breathe after you exercise. This is your body's way of supplying the muscle cells with the oxygen they need.

When yeast is used to make bread, alcoholic fermentation occurs. In making bread, CO_2 is given off by the yeast, causing the bread to rise. The holes that give bread its spongy texture are produced by the bubbles of CO_2 that are given off. The alcohol that is produced by the yeast evaporates in the hot oven during the baking process. Naturally, the process of fermentation is also used by the brewing industry where alcohol is the desired product! Many species of bacteria can also carry out alcoholic fermentation.

PHOTOSYNTHESIS

Respiration breaks down food and produces energy. The opposite process uses energy to produce food. Green plants and algae are able to use energy to produce food. This production of food is called **photosynthesis.** Photosynthesis uses the energy in light to produce carbohydrates from carbon dioxide and water. Green plants and algae are called **autotrophs** (self-feeders), because they are able to feed themselves by making their own food. All life on Earth depends on the food-making abilities of autotrophs. Four things are necessary for photosynthesis to occur: light, CO_2, H_2O, and chlorophyll (to trap the energy in light). The following equation describes the process of photosynthesis:

$$6CO_2 + 6H_2O + Energy \longrightarrow C_6H_{12}O_6 + 6O_2$$

Compare this equation to the equation for respiration on page 65. The photosynthesis equation is the reverse of the respiration equation.

Photosynthesis occurs in two phases, a **light phase** and a **dark phase.** Both phases occur in the chloroplasts of the cell. In

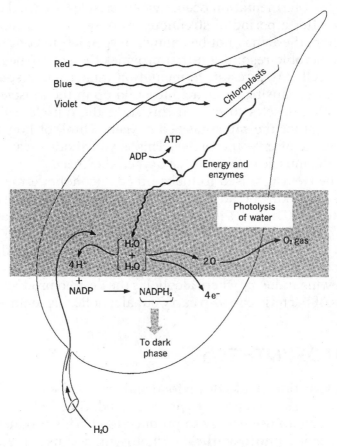

Figure 10-5. The light phase of photosynthesis.

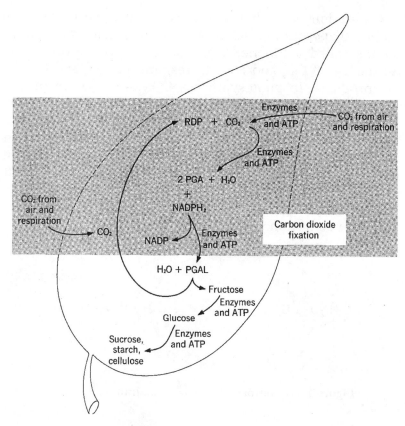

Figure 10-6. The dark phase of photosynthesis.

the light phase, light energy is absorbed by chlorophyll. The absorbed energy is used to split water into hydrogen and oxygen. The oxygen is a waste product and is given off by the cell, but the hydrogen is picked up and held in the cell by a **hydrogen acceptor, NADP,** changing it into NADPH$_2$. You can see in the chemical formula that a molecule of hydrogen has been picked up by NADP.

In the dark phase, ATP and NADPH$_2$ are used to build food from smaller organic molecules and CO_2. Glucose is usually the food that is produced. The dark phase is also called the **Calvin cycle.**

The rate at which photosynthesis occurs depends on several factors. For example, photosynthesis speeds up when light intensity increases and when temperature increases.

Respiration and photosynthesis are two very important processes that occur in cells. Both occur in a series of controlled reactions called a **biochemical pathway.** Both processes involve the use of CO_2 and H_2O. In respiration, CO_2 and H_2O are waste products. In photosynthesis, CO_2 and H_2O are the raw materials used. Respiration takes place in all living cells, but photosynthesis occurs only in cells that contain chlorophyll. Respiration breaks down sugars. Photosynthesis builds sugars. Glucose and oxygen are both the products of photosynthesis and the raw materials needed for respiration. These complementary processes allow cells to make energy-rich compounds by photosynthesis and to release the energy in those compounds by cellular respiration.

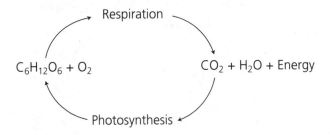

Figure 10-7. Photosynthesis and respiration.

QUESTIONS

Multiple Choice

1. The cell's main source of energy is   *a.* oxygen   *b.* protein   *c.* water   *d.* glucose.

2. The production of energy that uses oxygen is called   *a.* fermentation   *b.* aerobic respiration   *c.* anaerobic respiration   *d.* the Calvin cycle.

3. Fermentation produces either alcohol or   *a.* ATP   *b.* acetic acid   *c.* lactic acid   *d.* water.

4. Photosynthesis occurs in two phases,   *a.* ETC and glycolysis   *b.* light and dark   *c.* aerobic and anaerobic   *d.* Krebs and Calvin.

5. The breakdown of glucose into pyruvic acid is called   *a.* the Krebs cycle   *b.* photosynthesis   *c.* the ETC   *d.* glycolysis.

Matching

6. ATP   *a.* second stage of cellular respiration
7. ETC   *b.* series of electron carriers
8. citric acid cycle   *c.* dark phase
9. fermentation   *d.* stores energy
10. Calvin cycle   *e.* breakdown of glucose without oxygen

Fill In

11. In respiration, water and _____ are the waste products.
12. Light energy is absorbed by _____.
13. The Krebs cycle is also called the _____ _____ cycle.
14. _____ _____ fermentation can take place in humans.
15. When ATP releases its energy, it reverses back to _____.

Portfolio

Many science fiction stories tell about the effects of the death of green plants on Earth. Write your own short story that tells what would happen. You might like to make a series of drawings to "tell" your tale instead.

Chapter *11*

Cell Growth and Reproduction

In 1858, the German physician Rudolf Virchow stated that all cells come from other cells. This seems such a simple and obvious statement, but in its time it was quite revolutionary. Today, we know that cell division is the basis for the continuation of life. We know that when cells reach a certain size, they divide.

For many years, however, scientists did not know what caused a cell to begin to divide. Now we know that the outer surface of a cell grows more slowly than the volume of material inside the cell. Scientists call this the **surface-to-volume ratio.** When the surface of a cell is no longer large enough to let in the proper amounts of food and let out wastes, the cell divides. The two smaller cells have a larger ratio of surface to volume and are able to function better.

When a cell is ready to divide, it is called a **parent cell.** The two new cells produced as a result of cell division are called **daughter cells.** Each daughter cell should be identical to the parent cell it came from.

MITOSIS

Before a cell can divide, the nucleus must divide. The division of the nucleus is called **mitosis.** During mitosis, the nucleus divides to form two nuclei. Each nucleus has the same kinds and numbers of chromosomes as the parent or original cell. In a single-celled organism, mitosis results in two new organisms. In multicellular organisms, mitosis increases the number of cells. For example, if you cut your skin, cells begin to divide to fill in the cut. When the surface of your skin is healed, cell division stops.

In the nucleus, the chromosomes—which contain the encoded genetic information for the cell—must make a copy of themselves before cell division begins. The chromosomes are made up of a material called chromatin. During the early stages of cell division, the chromatin condenses and becomes visible. The copied, and now visible, chromosomes are called **chromatids.** Each pair of chromatids is held together by a **centromere.** Although the centromere can be located at any place on the chromatid, it is usually located near the middle. In order to make sure everything is copied exactly and divided evenly between daughter cells, a series of four phases occurs in the nucleus.

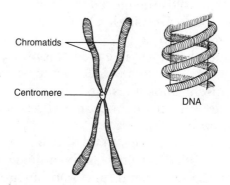

Chromatids

Centromere

DNA

Figure 11-1. Each pair of chromatids is held together by a centromere.

The first phase that occurs in mitosis is called **prophase.** During prophase, the longest of the phases, chromosomes are formed when chromatin coils. Also, both the membrane that surrounds the nucleus and the nucleolus break down. The **centrioles,** small dark bodies, separate and move toward opposite ends, or **poles,** of the cell. A structure made of microtubules forms between the centrioles. This structure is called a **spindle.**

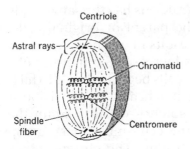

Figure 11-2. Prophase.

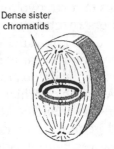

Figure 11-3. Metaphase.

The spindle helps position the chromosomes so that the number of chromosomes is equal in all new cells.

The second phase of mitosis is **metaphase.** During metaphase, the chromosomes line up along the **equator,** or the center of the cell.

Anaphase is the third phase of mitosis. During anaphase, the centromere that holds the paired chromosomes (sister chromatids) together splits. The single chromatids move away from the equator toward the poles.

The last stage of mitosis is called **telophase.** The spindle and centrioles disappear. The nuclear membrane forms around each new mass of chromosomes, and the nucleolus reappears. The chromosomes uncoil and become threadlike chromatin. Now the cell has two nuclei.

The cytoplasm now divides to form two new cells. Division of the cytoplasm is called **cytokinesis.** In animal cells, cytokinesis occurs when the cell membrane folds in until both sides

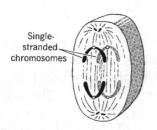

Figure 11-4. Anaphase.

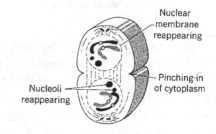

Figure 11-5. Telophase.

meet. Then the cell separates into two cells. In plant cells, a **cell plate** forms, dividing the cell into two new cells.

Mitosis and cytokinesis are part of a process called the **cell cycle.** The cell cycle also includes all the processes that occur between cytokinesis and the next mitotic division. This period of cell growth and development between divisions is known as **interphase.**

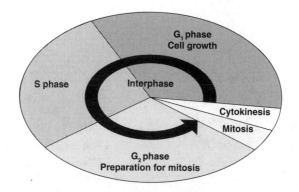

Figure 11-6. The relative time periods of various stages in the cell cycle.

Interphase has three divisions called the G1, S, and G2 phases. During the **G1 phase,** the cell grows in size, and the cell's organelles increase in number. During the **S phase,** the chromosomes replicate. And during the **G2 phase,** the cell continues growth and prepares for mitosis. Interphase, mitosis, and cytokinesis make up the cell cycle.

MEIOSIS

Mitosis produces two new cells that are identical to the parent cell except in size. Mitosis occurs in all cells, except those cells that are involved in sexual reproduction. The cells involved in sexual reproduction, cells that produce eggs or sperm, must reduce their number of chromosomes by half. This reduction is necessary so that when an egg and a sperm combine during fertilization, a full and correct number of chromosomes is found in the nucleus. The process of cell division that produces cells with half the number of chromosomes is known as **meiosis.**

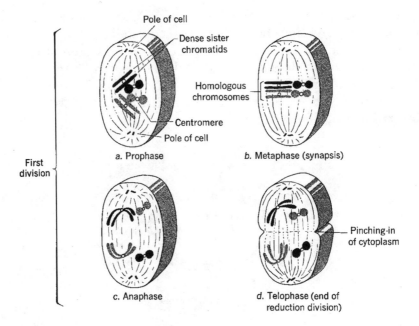

Figure 11-7. The first division of meiosis.

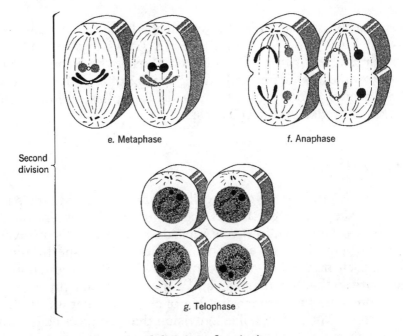

Figure 11-8. The second division of meiosis.

During meiosis, two nuclear divisions occur. The first division is called **meiosis I** and the second division **meiosis II.**

There are four phases in meiosis I. In **prophase I,** the DNA shortens into chromosomes, the spindle appears, and the nuclear membrane and nucleolus disappear. Also during this phase each chromosome lines up next to its similar chromosome. This pairing is called **synapsis.** The two chromosomes form a **tetrad** of four chromatids. In **metaphase I,** the tetrads line up along the equator. During **anaphase I,** the similar pairs separate, and each pair moves toward the poles. In **telophase I,** the cytoplasm splits to form two new daughter cells, and the chromosomes may uncoil.

In meiosis II, the two new daughter cells continue their nuclear division. During **prophase II,** the chromosomes shorten and become visible again. In **metaphase II,** the chromosomes move to positions along the equator. Each chromosome is made of two sister chromatids joined by a centromere. The centromere holds the chromatids onto the spindle. In **anaphase II,** the centromeres between the chromatids split, and the chromatids move toward the poles. During **telophase II,** the spindle dissolves, and the nuclear membrane reforms around the chromosomes. After meiosis II is complete, cytokinesis occurs. The result is four new daughter cells, each with half the number of chromosomes as the parent cell.

SEXUAL REPRODUCTION

Most animals and plants reproduce by **sexual reproduction.** In sexual reproduction, male and female sex cells join to form a fertilized egg, or **zygote.** The zygote divides again and again by mitosis and eventually becomes an adult organism.

Mature organisms produce sex cells—sperm or eggs. The production of sex cells, or gametes, is called **gametogenesis.** In females, the production of eggs is termed **oogenesis.** In males, the production of sperm is called **spermatogenesis.** The processes are different: in spermatogenesis, four functional sperm cells are produced, but during oogenesis, only one functional

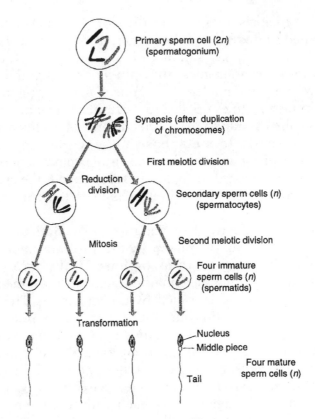

Figure 11-9. Formation of sperm (spermato-genesis).

egg cell is produced. Remember, however, that both sperm and egg cells have only one-half the number of chromosomes as do the other cells in the body.

ASEXUAL REPRODUCTION

Some animals and plants are produced by a single parent. This process is called **asexual reproduction.** There are several types of asexual reproduction. One type is **budding.** In budding, a daughter cell, or a group of cells, forms a bud from a parent cell or organism. The bud, which is smaller than the parent, simply falls free from the parent and begins to develop on its own. Some single-celled organisms split into two new cells that are equal in size. This process is called **binary fission.** Another type

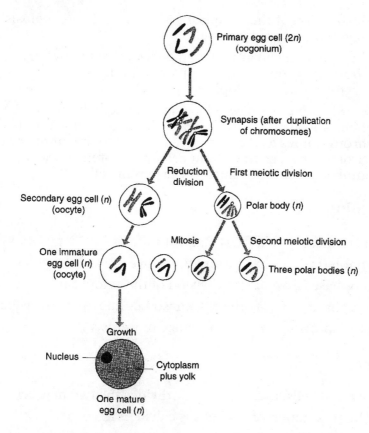

Figure 11-10. Formation of eggs (oogenesis).

of asexual reproduction is called **regeneration.** Regeneration allows an organism to replace lost or damaged structures. For example, a lobster grows or regenerates a new claw if it loses one in a fight.

QUESTIONS

Multiple Choice

1. Cells that are ready to divide are called   *a*. grandmother cells   *b*. parent cells   *c*. son cells   *d*. daughter cells.
2. In which of the following phases do chromosomes line up on the equator?   *a*. prophase   *b*. telophase   *c*. anaphase *d*. metaphase.

3. The division of the cytoplasm is   *a.* mitosis   *b.* meiosis
   c. cytokinesis   *d.* budding.

4. The period of cell growth and development between
   mitotic divisions is   *a.* prophase   *b.* interphase
   c. cytogenesis   *d.* telophase.

5. Mitosis produces new cells with   *a.* one-half the number
   of chromosomes as the parent cell   *b.* the same number of
   chromosomes as the parent cell   *c.* twice the number of
   chromosomes as the parent cell   *d.* one-fourth the
   number of chromosomes as the parent cell.

Matching

6. metaphase   *a.* first stage of mitosis
7. anaphase   *b.* chromatids move toward the poles
8. prophase   *c.* the division of the cytoplasm
9. telophase   *d.* the chromosomes line up on the equator
10. cytokinesis   *e.* the last stage of mitosis

Fill In

11. In plant cells, a _____ _____ divides the two new cells.
12. The production of sex cells is called _____.
13. The center of the cell where the chromosomes line up is
    called the _____.
14. Four chromatids combined make up a _____.
15. Cells may get the message to divide due to their _____
    ratio.

Portfolio

Gardeners take advantage of the ability of certain plants to reproduce asexually. For example, it is often possible to cut a
stem from a begonia or a geranium and root it in water. Do
some research in a library about reproducing plants by making
stem cuttings. Report your findings to your class. You can
make a display of some cuttings you made, or you can draw
what you observe as your cuttings develop over time.

UNIT FIVE
GENES AND HEREDITY

Chapter *12*

Genes, Chromosomes, and DNA

THE GREAT DIVERSITY of life on Earth has always been a cause of great wonder. What determines the scent of a rose, the color of a robin's feathers, or even the blobby shape of an ameba? What makes all elephants look slightly different while still having the same basic characteristics?

Within every cell is a code that contains all the information needed to produce a complete, complex organism. Cells also contain the important instructions that determine the characteristics of any offspring an organism produces.

CHROMOSOMES

In the early 1900s, the geneticist **Walter Sutton** identified **chromosomes** as the structures that transmit hereditary traits. His ideas became known as the **chromosome theory of inheritance.** He found that each chromosome controls a different set of characteristics.

Chromosomes contain different amounts of **DNA** or **deoxyribonucleic acid.** Chromosomes differ in size and shape, and perform different functions. Certain chromosomes determine the sex of an individual. These chromosomes are called the **sex chromosomes.** All the other chromosomes are called

autosomes. Each species has a particular number and kind of chromosomes in its cell or cells. For example, humans have 46 chromosomes, cats have 32, and horses have 64. However, the complexity of an organism does not depend on the number of chromosomes it has.

GENES

Over time, scientists realized that there were many more traits than there are chromosomes. **Thomas Hunt Morgan** discovered that chromosomes were probably made up of many smaller sections called genes. Morgan thought that it was the genes that actually determined the expression of a particular trait. Morgan's ideas were called the **gene theory of inheritance.**

Genes are the basic units of heredity. They are segments of DNA that contain the message, or code, for a particular trait. Actually, genes are not located "on" chromosomes. Genes, and the proteins that are bound to the genes, *are* the chromosomes. Cells control which genes are active and which gene products appear. Scientists are able to construct **chromosome maps** that show the relative position of genes on a chromosome. Today, scientists are working on a map of all the genes in a human, an enormous and important task.

DNA: HISTORY AND STRUCTURE

Because genes are vital to a cell's functions, each new cell must have an exact copy of the genes found in its parent or parents' cells.

DNA was discovered by **Johann Miescher** in 1869. Some scientists thought that DNA could make up genes. By the 1920s, scientists had analyzed chromosomes and found that they contained proteins and RNA, as well as DNA. Many years passed before scientists completely understood that DNA is the most important molecule that makes up genes. Even more years passed before scientists realized that DNA, in the form of genes, controls the production of every single protein in a cell. And it is the cell's proteins that control all of the processes in a cell.

In 1928, **Frederick Griffith** performed a very important experiment trying to isolate the "**transforming factor**" that changed nondeadly bacteria into deadly bacteria. Griffith had transformed one strain of bacteria into another. But his experiment raised as many questions in Griffith's mind as it answered. He did not know what had caused the change in the bacteria. Was the transforming factor a protein, or was it DNA?

In 1944, **Oswald T. Avery** and a team of scientists at Rockefeller University decided to repeat Griffith's experiment in an attempt to discover the identity of the "transforming factor." They determined that it was DNA that transformed the harmless bacteria into disease-producing bacteria. The DNA had changed the characteristics of the harmless strain. Still, many scientists argued against the conclusions of Avery and his team.

However, proof of the importance of DNA came in 1952 from the so-called blender experiment conducted by **Alfred Hershey** and **Martha Chase.** They used radioactive isotopes to label DNA and proteins. Their results clearly indicated that DNA was the transforming factor and therefore the carrier of genetic information.

THE STRUCTURE OF DNA

Before scientists could understand how genetic information was passed from one generation to another, they had to determine the structure of DNA. By 1900, the basic chemistry of DNA

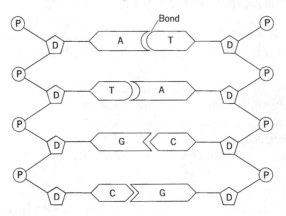

Figure 12-1. Structure of DNA.

had been worked out. Final pieces of the puzzle came together in the 1950s with the work of **Linus Pauling** on helical structures (structures that look like a circular staircase), data by **Erwin Chargaff** on the complementary pairing of the bases (see below), and the X-ray diffraction photographs of DNA taken by **Rosalind Franklin.**

Finally in 1953, **James Watson,** an American geneticist, and **Francis Crick,** a graduate student in physics at Cambridge University, were able to construct a model of DNA that showed its structure. They finally assembled all the pieces of the puzzle that had been accumulating for 80 years. The discovery of the structure of DNA may be the most important scientific discovery of the twentieth century.

CHEMICALS THAT TRANSMIT DATA

DNA is made of smaller units called **nucleotides.** Each nucleotide is composed of a sugar molecule, a phosphate molecule, and a nitrogen base. There are four different bases in a molecule of DNA: cytosine, guanine, thymine, or adenine (Figure 12–2). DNA looks like a twisted ladder. This special shape is known as a **double helix** (Figure 12–3). The sides of the ladder are made up of joined sugar-phosphate molecules. The "rungs" are made from a pair of bases held together by hydrogen bonds. Watson and Crick determined that the bases joined in a certain way. They found that adenine always paired with thymine and that cytosine always paired with guanine. The base pairs are called **complementary bases.**

DNA REPLICATION

It is the very structure of DNA that enables it to make exact copies of itself. This process is called **DNA replication.** In order for a strand of DNA to copy itself, it must first unwind. The weak hydrogen bonds between the base pairs break, and the two strands (sides of the ladder) separate. This allows free DNA nucleotides present in the nucleus of a cell to match up with the

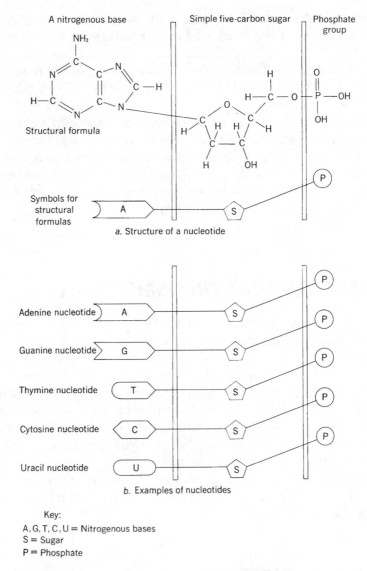

Figure 12-2. Structure and examples of nucleotides.

complementary bases of the two separate strands. The original strands act as a **template,** or a pattern, for the construction of two new DNA strands. Each new strand consists of one side of the "old" ladder and a completely new strand formed from materials floating in the nucleus of the cell.

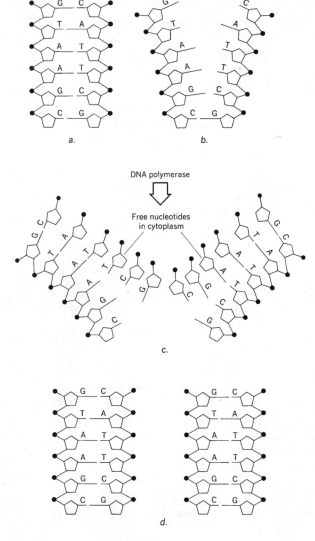

Figure 12-3. DNA replication.

PROTEIN SYNTHESIS

DNA has all the stored information needed to determine the sequence of amino acids in proteins. The building of proteins, called **protein synthesis,** actually directs the formation of protein molecules. This assembly takes place outside the nucleus on the ribosomes. Since DNA remains in the nucleus, how does the information stored in the DNA get to the ribosomes? DNA sends instructions for building proteins to the ribosomes in the form of messenger RNA (mRNA). Like DNA, RNA is made up of a chain of nucleotides. However, there are several important differences between DNA and RNA. RNA consists of only a single strand of nucleotides. The sugar in RNA differs from the

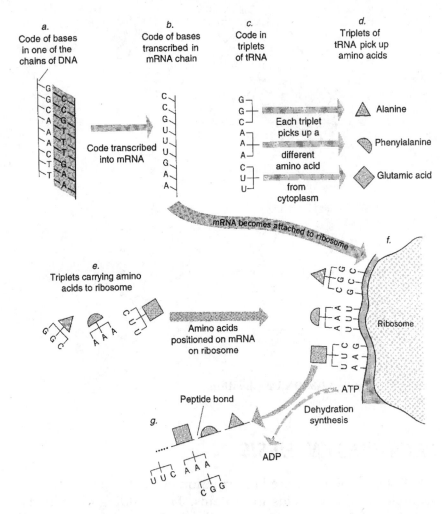

Figure 12-4. Protein synthesis.

sugar in DNA. The most important difference is in the four bases that make up a strand of RNA. Remember that DNA is made up of adenine, guanine, thymine, and cytosine. Adenine, guanine, and cytosine also are found in RNA. But instead of thymine, RNA contains the base uracil. In RNA, cytosine bonds to guanine, and adenine bonds to uracil.

The process by which the DNA message is copied onto a strand of mRNA is called **transcription.** Messenger RNA carries the instructions that direct the assembly of a specific protein to a designated area on a ribosome. The instructions are carried in a series of sequences of three bases called **codons.**

Once the message has reached the ribosome, the protein is ready to be assembled. The process of building the protein from the mRNA instructions is called **translation.** Transfer RNA (tRNA) and ribosomal RNA (rRNA) are involved in translation. **Transfer RNA** is responsible for carrying amino acids (the building blocks of proteins) to the ribosome so they can be linked in the specific order that makes up a single protein. One

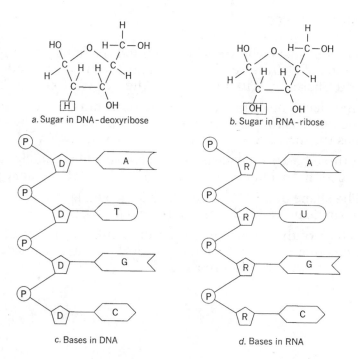

Figure 12-5. DNA and RNA structure.

end of the transfer RNA carries a three-base sequence called an **anticodon,** which will match up with a particular codon on the mRNA. **Ribosomal RNA** helps bond the amino acids together to form the final protein. Protein synthesis stops when a termination codon is read on the mRNA. After synthesis is stopped, the new protein is released from the ribosome for use throughout the cell (Figure 12–4).

Every protein a cell produces is directly related to the order of the bases in the mRNA, and the sequence of bases of mRNA is complementary to the sequence of bases of the DNA. In other words, DNA is the ultimate blueprint for the formation of proteins. In 1957, Francis Crick used this logic to propose what is now known as the **central dogma.** The central dogma is shown as: DNA→ RNA→ Protein. It illustrates the flow of information needed to build proteins.

QUESTIONS

Multiple Choice

1. The scientists who were able to build the first correct model of DNA were   *a.* Beadle and Tatum   *b.* Hershey and Chase   *c.* Wilkins and Franklin   *d.* Watson and Crick.

2. In order to build proteins, instructions must be sent from the nucleus to the ribosomes. Which of the following carries the message?   *a.* tRNA   *b.* mRNA   *c.* rRNA   *d.* DNA

3. A three-base sequence on an mRNA strand is called the   *a.* codon   *b.* anticodon   *c.* triplet   *d.* complementary pair.

4. The shape of a DNA molecule is a   *a.* single strand   *b.* single helix   *c.* double helix   *d.* cloverleaf.

5. Which of the following did Sutton identify as the structures that transmit hereditary traits?   *a.* genes   *b.* DNA   *c.* RNA   *d.* chromosomes

Matching

6. replication
7. translation
8. transcription
9. complementary pairs
10. chromosome mapping

a. assembling proteins from mRNA code

b. making copies of DNA

c. A-T and G-C

d. production of mRNA from DNA

e. position of genes on a chromosome

Fill In

11. The tRNAs _____ must match up with the codon on the mRNA in order to have the correct sequence of amino acids.
12. DNA was discovered by _____ in 1869.
13. Bases that pair with each other are called _____ pairs.
14. _____ are the basic units of heredity.
15. The building of proteins is called protein _____.

Free Response

16. Describe the process of protein synthesis.
17. How does DNA make a copy of itself?
18. Describe the scientific discoveries that led to the Watson and Crick model of DNA.
19. Why are chromosome maps important in modern genetics?
20. Describe the structure of a nucleotide.

Chapter *13*
Basics of Heredity

HAVE YOU EVER been at a school event and noticed the resemblance between members of the same family? The members of some families have long slender faces, while the members of other families have round faces. Other physical traits are easily identified too. The traits a family shows are passed down from one generation to the next. For many years, people believed that traits were passed to the next generation through the blood. Now we know that the chromosomes carry traits from one generation to the next.

Heredity is the passing of traits from one generation to the next generation. Receiving these traits from parents is called **inheritance.** An Austrian monk, **Gregor Johann Mendel,** was the first researcher to describe the inheritance of traits. In the mid-1850s he studied inheritance by crossing or breeding pea plants in his garden. Pea plants were perfect for his experiments because they are easy and quick to grow and they show several clearly contrasting traits. His seven-year study of inheritance in pea plants allowed him to collect data, analyze them, and eventually explain the basic principles of heredity. **Genetics** is the study of heredity. Mendel is referred to as the **"founder of genetics"** because of his important contributions to the study of heredity.

MENDEL'S WORK

Mendel studied seven pairs of contrasting traits that are easily observed in pea plants. For example, he crossed plants and recorded their stem lengths, the color and texture of seeds, and types of seed pods. Mendel predicted the possible outcome of his crosses. He used **probability,** or the mathematics of chance, to make his predictions. In some of his experiments he used **pure lines** of traits. These pure lines refer to a plant and its offspring that show a particular trait generation after generation. Pure lines were achieved by allowing the plants to self-fertilize for several generations. When two different pure lines were crossed, a **hybrid** plant was produced. This hybrid generation of plants is called the **first filial generation** or F_1 **generation.**

In one of his experiments, Mendel crossed a pure tall pea plant with a pure short pea plant. Mendel suspected that medium-sized plants would be produced. However, Mendel found that only tall plants appeared in the F_1 generation. When Mendel experimented with similar crosses for all the pure lines, he got the same result. One trait was outwardly shown, and the other trait seemed to disappear. Mendel then allowed hybrid

Table 13-1.   Crosses of Parents Having Contrasting Characters

Trait	P_1 Generation (Pure Types)	F_1 Generation (Hybrids)
Length of stem	Long stem X short stem	All long (100%)
Type of seed coat	Smooth seed X wrinkled seed	All smooth (100%)
Color of seed coat	Colored X white	All colored (100%)
Form of pea pod	Thick X thin	All thick (100%)
Position of flower	Side of stem X end of stem	All side of stem (100%)
Color of cotyledons	Yellow X green	All yellow (100%)
Color of pod	Green X yellow	All green (100%)

plants from the F_1 generation to self-fertilize. In the next generation of plants, the **F_2 generation,** Mendel was surprised to find that both pure-line traits appeared. That is, some of the new plants were tall, other plants were short. This meant that the trait that appeared to be lost in the F_1 generation had reappeared in the F_2 generation.

MENDEL'S LAWS

To explain these unexpected results, Mendel hypothesized that **factors** controlled heredity, and that these factors occurred in contrasting pairs. For example, one factor controlled the trait for tall pea plants, and the opposite factor controlled the trait for short pea plants. In analyzing his results, Mendel concluded that some traits were stronger or more assertive than others. He called the strong traits **dominant** traits and the weaker traits **recessive** traits. Mendel's **Law of Dominance** states that when a (pure) dominant and a (pure) recessive trait are present, the dominant trait will show in the hybrid organisms. But, remember, the recessive trait does not disappear. It shows up again in the F_2 generation.

Today we know that there are times when neither trait in a contrasting pair is dominant over the other. This condition is called **incomplete dominance.** When this situation occurs, two opposite traits are combined and a blending of traits appears to happen. For example, if a flower is incompletely dominant for color, and a red flower is crossed with a white flower, a pink flower results. However, in the F_2 generation, the original traits reappear. When two pink (hybrid) carnations are crossed, one-quarter of the F_2 generation will be white, one-quarter will be red, and one-half will be pink.

From his data gathered in the F_2 generations, Mendel proposed the **Law of Segregation.** This principle said that each individual carries two factors for each trait. This pair of factors was inherited from the parents, with each parent contributing one factor. The individual's factors segregate during meiosis so that one factor ends up in each sperm or egg cell.

Mendel even crossed plants that were different in two traits,

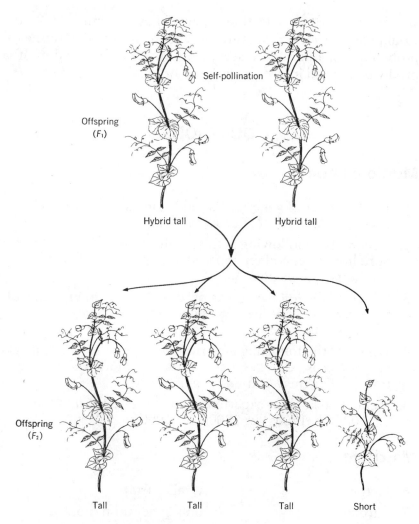

Figure 13-1. Mendel's experiment.

such as seed color and plant height. An analysis of his results showed that factors for different traits were inherited independently. Mendel called this the **Law of Independent Assortment.**

Mendel had unknowingly begun the groundwork for the modern science of genetics. He correctly hypothesized, and performed, accurate experiments that supported his theories. Unfortunately, most of Mendel's work was not discovered until 1900, several years after his death. He never knew the impact his work would have on society. We now use his theories to determine the probability of inherited disease and birth defects, and even to produce new and improved varieties of plants and animals.

QUESTIONS

Multiple Choice

1. The "founder of genetics" is   *a.* Thomas Hunt Morgan   *b.* Gregor Mendel   *c.* Leeuwenhoek   *d.* Aristotle.

2. Which of the following is the result of the crossing of two pure lines?   *a.* a phenotype   *b.* an allele   *c.* a hybrid   *d.* a homozygous pair

3. Which of the following was used by Mendel in his study of heredity?   *a.* fruit flies   *b.* guinea pigs   *c.* roses   *d.* pea plants

4. Which of the following generations appears if two F_1 members are crossed?   *a.* P generation   *b.* another F_1 generation   *c.* F_2 generation   *d.* F_3 generation

5. Mendel's pea plant experiments were conducted in the   *a.* 1600s   *b.* 1700s   *c.* 1800s   *d.* 1900s.

Matching

6. factors
7. pure line
8. recessive
9. F_1
10. independent assortmentt

a. weaker trait

b. first-generation cross

c. alike generation after generation

d. Mendel's trait carriers

e. factors for different traits are inherited separately

Fill In

11. According to Mendel, a strong trait is called a _____ trait.

12. Mendel's Law of _____ stated that factors separate when the sex cells are formed.

13. If neither trait is stronger, the condition is called _____ _____.

14. Mendel used the mathematics of chance, or _____, to make predictions about his genetic crosses.

15. Mendel hypothesized that each individual carries _____ factors for each trait.

Portfolio

1. Learn more about the traits Mendel studied in pea plants. Make a drawing that shows the results of several of his crosses. You might like to make a bulletin board display for your class.

2. You might be able to duplicate some of the crosses that Mendel made. Remember, you must use pure strains of peas available from a laboratory supply house for your crosses. Hybrid seeds will not work in this kind of investigation. Record your findings in your portfolio.

Chapter 14
Genetics

DID YOU EVER watch fruit flies swarming around a piece of ripe fruit? Fruit flies are one of the most common organisms used to study how traits are passed from one generation to another. Why would scientists choose this organism to study? They are small and easy to keep in a laboratory; they reproduce and develop quickly; they are inexpensive to acquire and maintain; and their inherited traits are easy to observe. Keep these reasons in mind as you read about several classic experiments and the scientists who worked hard to understand how traits are transmitted.

THE SCIENCE OF HEREDITY

Heredity is the passing of traits from one generation to the next generation. **Genetics** is the study of heredity. A **geneticist** is a scientist who studies genetics. In the years since the discovery of Mendel's work with pea plants and heredity, geneticists have developed a special vocabulary to describe heredity.

Mendel's factors are now called **genes.** A gene is a section of DNA on a chromosome that controls the production of a specific trait or protein. Since chromosomes occur in pairs, genes occur in pairs. Each half of the gene pair is called an **allele.** Geneticists use letters to represent alleles. For example, a capital T could be used to represent a gene for tall, and a small t could be

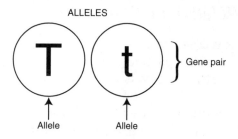

Figure 14-1. Two alleles: one dominant, one recessive.

used to represent a gene for short. **Homozygous** is the term used to describe an organism that has inherited two identical alleles and is pure for that trait. In our example, they would be *TT* (pure tall) or *tt* (pure short). **Heterozygous** describes an organism that has two different alleles for a trait. In our example, that would be *Tt*. Hybrids are heterozygous.

GENOTYPES AND PHENOTYPES

Genotype describes the actual genetic makeup of an individual organism. It describes the combination of alleles. For example, if tall *(T)* is dominant over short *(t)*, a tall pea plant could be *TT* or *Tt*. Either combination of genes will produce a tall plant. A short plant would be *tt*. Only this combination of two genes for shortness will produce a small plant.

A genotype allows us to recognize unexpressed traits (recessive traits that are present but not shown) that are inherited from the parents. When a couple undergoes genetic testing prior to having a baby, a scientist must determine the couple's genotype. This allows the parents to find out if they have genes for diseases such as sickle-cell anemia, hemophilia, or Tay-Sachs.

Phenotype describes the outward expression of the genotype. These are the traits that we see, such as hair color, eye color, and height. A phenotype does not, in all cases, tell us the genotype of the individual. A phenotype can, in some instances, be affected by the environment. The dark ears of a Himalayan rabbit show up as a result of their cool temperature in relation to the rest of the rabbit's body.

USING A PUNNETT SQUARE

Punnett squares are diagrams geneticists use to show the possible genotypes of a particular genetic cross. Punnett squares allow us to visualize the possible outcomes of a cross. When studying a cross involving only one trait (height of pea plants, for example), a Punnett square of four boxes is used. When studying a cross that involves two traits, the Punnett square must have 16 boxes.

An example of both a one-trait and a two-trait cross will demonstrate how Punnett squares are used. Suppose a homozygous tall plant and a short plant are crossed. Tall is dominant to short. We are looking only at one trait—height. We can represent our pure tall by using the letters *TT*. Our short plant, also homozygous, will be represented as *tt*. You can see the possible outcomes of this cross if you use a Punnett square.

We separated the alleles according to Mendel's law of segregation. You will notice in our example that all the offspring will be tall (the phenotype). Even though each tall plant is carrying a recessive gene for shortness (genotype), that gene for shortness is not being expressed because tall is dominant. Our phenotype for the F_1 is four tall. The genotype is four heterozygous tall *(Tt)*. Keep in mind that the Punnett square gives the possible results of a cross. In certain crosses, the results shown in a Punnett square may not become apparent until a large number of offspring are produced. The larger the number of offspring, the greater the probability that the results will approach the theoretical results shown in the Punnett square.

In a cross involving two traits, the Punnett square is larger. Let's assume the tall trait is dominant to short and green pods

Figure 14-2. Punnett squares are used to show possible crosses.

are dominant to yellow pods. In this example, we are examining two traits, plant height and the color of the pea pods. What is the outcome of the cross between a homozygous tall, heterozygous green-pod plant and a homozygous short, homozygous yellow-pod plant? The key below shows the possible alleles.

Key:
TT	homozygous tall	*GG*	homozygous green
Tt	heterozygous tall	*Gg*	heterozygous green
tt	homozygous short	*gg*	homozygous yellow

In this example, the genotype of the two plants will be *TTGg* x *ttgg*. Remember, since we are studying two traits, the Punnett square will have 16 boxes. To use Mendel's laws correctly, you must make sure to segregate the alleles properly. To do this, use the following number combinations: 1–3, 1–4, 2–3, 2–4. Number each allele on both sides of the cross. Fill in the outer edges of the Punnett square with the four possible genetic combinations of each parent. The Punnett square gives you all possible genotypes of the cross.

MENDEL'S WORK REDISCOVERED

From the time of Mendel until the early twentieth century, the importance of his work was unrecognized by the scientific community. In the early 1900s, scientists saw a relationship between Mendel's data and their own studies of chromosomes. Chromosomes seemed to follow the patterns uncovered by Mendel. Scientists soon realized that Mendel's factors (genes) were located on the chromosomes. Different genes were found on different chromosomes. This idea became known as the **chromosome theory**—the theory that formed the basis for the science of genetics.

Thomas Hunt Morgan, a researcher at Columbia University, changed genetic research when he chose fruit flies as his experimental organism. Morgan and the other geneticists at Columbia were known as the "fly squad." Fruit flies, *Drosophila*, were perfect for genetic experiments. They have only four pairs of chromosomes, and the salivary glands have giant chromosomes that are easy to see. It is also easy to tell the differences

between male and female flies. It
was very helpful that the matura-
tion and the reproductive cycle
for fruit flies are short. Eggs develop into adults in about two
weeks. Because of their short life
cycle, more data could be collected. Finally, because of their
small size, thousands of experimental flies can be grown in a
few glass jars.

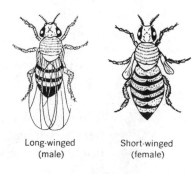

Long-winged   Short-winged
(male)   (female)

**Figure 14-3. Fruit flies are often
used by scientists who study the
inheritance of traits.**

From his research, Morgan
reasoned that genes determine
specific traits, and that many genes are located on a single chromosome. Morgan also discovered that some traits in the fruit fly
are carried on the sex chromosomes. (Sex chromosomes determine whether an organism is male or female.) In many organisms, such as fruit flies and humans, a male contains a pair of
sex chromosomes that do not match, an X and a Y. A female contains a matched pair of XX chromosomes. In some cases, only
the X chromosome carries a trait, not the Y chromosome. As an
example, the trait for eye color in fruit flies is found on the X
chromosome. Genes for traits located on the sex chromosomes
are called **sex-linked** traits.

Mendel and other early geneticists provided the groundwork for our modern study of heredity. Today, we are even able
to understand the chemical basis of heredity and use that information to study human inheritance. The application of genetics
has become an important scientific tool in our constant search
to improve our quality of life.

QUESTIONS

Multiple Choice

1. Traits that are carried on the sex chromosomes are called
   a. dominant traits   *b*. pure traits   *c*. sex-linked traits
   d. recessive traits.
2. The study of heredity is called   *a*. taxonomy   *b*. biology
   c. genetics   *d*. cytology.

3. Thomas Hunt Morgan and his fellow researchers were known as the   *a.* first true geneticists   *b.* gene team   *c.* fly squad   *d.* chromosome crusaders.

4. In doing a two-trait cross, a Punnett square must be made of   *a.* 4 boxes   *b.* 8 boxes   *c.* 12 boxes   *d.* 16 boxes.

5. In describing someone's eye color, you are identifying the *a.* phenotype   *b.* genotype   *c.* allelic frequency *d.* genetic variation.

Matching

6. phenotype
7. genotype
8. homozygous
9. heterozygous
10. Punnett square

a. identical alleles
b. helps you find genotypes
c. outward appearance
d. actual genetic makeup
e. different alleles

Fill In

11. An XY chromosome pair in a human would be a _____.
12. If green *(G)* is dominant to yellow *(g),* a heterozygous green would be shown as _____.
13. Thomas Hunt Morgan used _____ as his experimental organisms in his genetic research.
14. One-half of a gene pair is called an _____.
15. If we look at a plant and see its height and flower color, we are observing its _____.

Portfolio

Mendel's work remained out of sight for many years. Could this happen today? Write a short science fiction story about the work of a scientist whose study of life on Mars was recently discovered in the attic of an old house.

Chapter *15*

Genetic Variation

SOME YEARS AGO, a famous seed company offered a ten-thousand-dollar reward to the person who grew the first white marigold—a flower that is normally yellow to red in color. The reward was claimed when a keen-eyed observer noticed a white flower in a bed of pale yellow marigolds. You may notice the same kind of phenomenon if you walk through a botanical garden and see an odd-color flower growing in a bed of normal-color flowers. What causes this kind of change to occur?

MUTATIONS

Many times an unexpected change occurs in genetic information due to variations in the genetic material. These changes are called **mutations. Hugo de Vries,** in 1903, was the first scientist to refer to abrupt changes in genetic material as mutations. He called these new forms **mutants. Thomas Hunt Morgan** provided evidence of mutations with his famous experiments that examined inheritance in fruit flies in the early 1900s. During the 1920s, **Hermann Muller** studied mutations in greater depth and provided data that showed that mutations could be caused by X rays.

Mutations can occur randomly and accidentally in any cell. If a mutation occurs in a body cell other than a sperm or an egg çell, the changes are passed on to all the cells that come from it as a result of mitosis. This is called a **somatic mutation.** These changes are not passed on to the offspring of the organism.

Mutations can also occur in the genetic material of the reproductive cells—sperm and eggs. These are known as **germ cell mutations.** This type of mutation can be passed to the zygote and show up in the offspring.

Mutations are caused by mutagens. **Mutagens** are substances or conditions that cause an error in replication of genetic material. Some chemicals are mutagens. Radiation also is a mutagen. In some cases, even increases in temperature can produce mutations.

Mutations are usually classified into two categories: chromosomal mutations and gene mutations. If the structure of a chromosome or the number of chromosomes is changed, it is called a **chromosomal mutation.** Chromosomal mutations can cause changes in the cell and the whole organism. **Gene mutations** occur in individual genes and involve changes in the DNA code. If the order of bases in a DNA strand is changed, the gene with the change may now code for a different protein. The effects produced as a result of a gene mutation may be profound.

CHROMOSOMAL MUTATIONS

Chromosomal mutations can result from a variety of problems. Most chromosomal mutations occur during cellular divisions. Sometimes the wrong number of chromosomes ends up in a cell due to **nondisjunction.** Nondisjunction occurs when homologous pairs of chromosomes fail to separate during meiosis. When this happens, one of the resulting gametes has an extra chromosome, while the other gamete lacks a chromosome. **Trisomy** is the condition caused by an extra chromosome in a cell. **Monosomy** results when a cell is missing a chromosome. Monosomy is usually more severe, since all the genetic material on the missing chromosome is lost.

Down syndrome is a disorder caused by nondisjunction. People with Down syndrome have an extra number 21 chromosome (trisomy-21). People with Down syndrome are usually

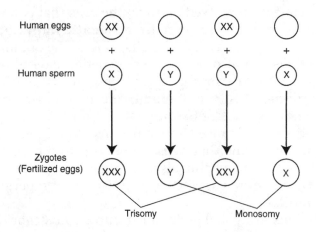

Figure 15-1. Nondisjunction of sex chromosomes.

short in stature and may have a variety of physical and mental disorders.

Klinefelter and **Turner** syndromes are caused by nondisjunction in the sex chromosomes. Males with Klinefelter syndrome have an extra X chromosome (XXY). Turner syndrome individuals are females with an XO genotype; they lack a second X chromosome.

In the production of the sex cells, two divisions must occur during meiosis. Sometimes a cell does not undergo the second division. The resulting gametes have diploid numbers instead of haploid numbers. Organisms produced from these gametes have an extra set of chromosomes. They are called **polyploid.** Even though this condition is common and harmless in many plants, it is usually lethal when it occurs in animals.

Structural changes in the bases that make up a chromosome can also occur. These structural changes include deletion, translocation, and inversion. **Deletion** occurs when a piece of DNA breaks off and does not rejoin the chromosome. All of the genetic information on this section of chromosome is permanently lost. **Translocation** occurs when a fragment from one chromosome attaches to another chromosome. In this case, genetic information is lost from one chromosome and new genetic information is added to the second chromosome. Sometimes

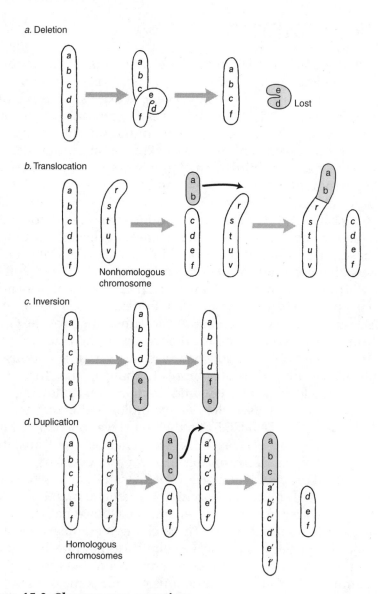

Figure 15-2. Chromosome mutations.

a section of a chromosome detaches and rejoins, but in an up-side-down position. This is called an **inversion.** Even though none of the genetic information is lost, the correct order of the bases in the original chromosome has been changed.

GENE MUTATIONS

Gene mutations may involve a change in a large section of DNA or a change in a single base in a nucleotide. Either change may be enough to affect the way a gene is expressed. Radiation, certain chemicals, and even some viruses can cause a gene mutation. **Albinism** is caused by a gene mutation. Albino animals have white hair and pink eyes and skin. In an animal that expresses the trait for albinism, melanin is not produced. Melanin is a pigment that affects skin, eye, and hair color.

If the mutation involves only one base in a nucleotide, it is called a **point mutation.** Point mutations can include the addition of a new base, the deletion of a base, or the substitution of one base for another. If a substitution occurs, the order of the bases is changed, and the genetic code is altered. This results in the insertion of an incorrect amino acid at a particular location as a new protein is being assembled. For example, if the original DNA sequence was CGA, and an A was substituted for the G, the mRNA message would code for the wrong amino acid.

When deletions or additions of bases occur, a **frameshift mutation** results. At the point of change, the three-base sequence (codon) is altered, and each codon following that sequence is altered as well. For example, if the sequence of the DNA was CGG/TAG/CCT and the first G was deleted, the new sequence would be CGT/AGC/CT. This changes the translation of the whole message, resulting in the production of a different protein. This could change an entire trait in an organism.

Even though mutations occur, most genetic material passes unchanged from generation to generation. Occasionally, changes in the genetic material occur. These changes are chance events. They may be harmful, they may be helpful, or they may have no effect on the traits passed on, since several RNA triplets code for the same amino acid. Certainly mutations provide some of the variation that is important to the process of evolution. Indeed, it is the variations within a population that are acted upon by natural selection. According to Darwin's theory of evolution, organisms compete to exist. Individuals with traits that aid survival live to reproduce and pass on those traits. Genes, and the random mutations that occur in them, provide the variations in traits. Over time, evolution occurs—the accumulation of genetic variations that help organisms survive and adapt to changes in their environment.

QUESTIONS

Multiple Choice

1. Down syndrome is caused by   *a.* translocation
   b. nondisjunction   *c.* a frameshift   *d.* none of the above.
2. The substances or conditions that cause mutations are
   called   *a.* mutants   *b.* polyploids   *c.* mutagens
   d. monoploids.
3. Which of the following scientists first used the term *mutation* to describe unexpected genetic changes?
   a. Hugo de Vries   *b.* James Watson   *c.* Hermann Muller
   d. Thomas Hunt Morgan
4. When homologous pairs of chromosomes fail to separate
   during meiosis, it is called   *a.* translocation   *b.* inversion
   c. duplication   *d.* nondisjunction.
5. Which of the following could be a mutagen?   *a.* a rise in
   temperature   *b.* radiation   *c.* certain chemicals   *d.* all of
   the above

Matching

6. translocation
7. inversion
8. deletion
9. point mutation
10. frameshift

   a. breakage, reattachment in
   a reverse orientation
   b. breakage, reattachment to
   another chromosome
   c. a piece of chromosome
   breaks off and is lost
   d. substitution at a single
   base site
   e. result of an addition or
   deletion at a base site

Fill In

11. _____ mutations take place in body cells and are passed
   on only to daughter cells.
12. Turner syndrome is caused by nondisjunction in the _____
   chromosomes.

13. If a mutation involves only one _____ in a nucleotide, it is called a point mutation.

14. An _____ mutation causes the correct order of the genetic information to be altered, even though no genetic information is lost.

15. A _____ is an organism that contains a mutation.

Free Response

16. Explain the differences between chromosomal and gene mutations.

17. Describe three types of chromosomal mutations.

18. What is albinism and what causes it?

19. What are mutagens? Name two types of mutagens.

20. Compare the conditions of trisomy and monosomy.

Chapter 16

Biotechnology

ONE OF THE new and revolutionary areas of biological research is the field of **biotechnology**. Biotechnology is "applied biological science," such as the use of the genetic material in living organisms to help make useful products or to solve medical problems. The use of biotechnology has affected the ways we diagnose and even prevent human diseases, many practices in agriculture, and even areas of criminal investigations.

Biotechnology is a combination of several different technologies. Even though biotechnology is a new word, the concept behind biotechnology is very old. Throughout history, people selected strains of bacteria and yeasts that were useful in producing certain food products. For example, they used yeast to make bread. Yeast is a microscopic organism related to mushrooms and the fungi that cause diseases such as athlete's foot and ringworm. Various kinds of bacteria were used to produce cheeses and yogurt. Bacteria are also living organisms. By making observations, and through trial and error, these selections could certainly be considered early uses of biotechnology.

GENETIC ENGINEERING

Scientists use biotechnology in much more sophisticated ways today. Scientists can actually use microorganisms to make many biologically important substances. For example, most genetic research is done at the molecular level. Scientists are now able to manipulate the genes of living organisms. This technique is usually called **genetic engineering.** Many times, genes are actually moved from one DNA molecule and inserted into another. The new DNA molecule is called **recombinant DNA.**

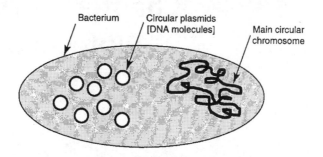

Figure 16-1. Bacterium with plasmids.

Scientists are able to combine two different DNA fragments through the use of **restriction enzymes.** Using restriction enzymes like a pair of chemical scissors, scientists cut a strand of DNA at a particular point in the sequence of bases. The point where the sequence is cut by the restriction enzyme is called the **restriction site.** Scientists then remove the fragment of DNA that contains a specific gene and insert that section into a new DNA molecule. The DNA fragment cannot function by itself; it must be inserted into the DNA of an organism. After insertion, the open areas of the DNA have to be closed. **Ligases** are enzymes used to join the pieces of DNA.

The process by which a section of DNA from one organism is inserted into the DNA of another organism is called **gene splicing.** It is easiest to insert DNA fragments into the DNA material present in bacterial cells. (Recall that bacterial cells lack a membrane surrounding the cell's DNA.) **Plasmids,** or circular pieces of DNA in bacteria, usually serve as the site of insertion for sections of DNA. When the plasmid replicates, it copies the new recombinant DNA. Because bacteria reproduce very quickly, many copies of the recombinant gene can be made in a short time. **Cloning** is asexual reproduction that produces identical copies of the DNA.

FRAGMENT SEPARATION AND REPLICATION

In order to manipulate DNA, scientists need to study the individual fragments of DNA they are working with. **Gel electrophoresis** is a method used to separate DNA fragments. This technique uses an agarose gel and an electric current. DNA is placed in the gel, and an electric current is run across the gel. Because DNA fragments are negatively charged, they move toward the positively charged areas in the agar. Small fragments

Intact Plasmid

1. Restriction enzymes make cuts on both DNA strands of a plasmid.

Broken Plasmid

2. The free ends of the plasmid have bases that will pair with complementary exposed bases on any other DNA strand.

DNA fragment

3. A DNA fragment containing useful genes is chosen to be inserted into the plasmid.

Recombinant DNA

4. The two different DNA molecules can base pair at their sticky ends; they then can be sealed together by DNA ligase.

Figure 16-2. Recombinant DNA.

of DNA move faster than larger fragments. Thus, based on its rate of migration, the size of the DNA fragment can be calculated.

It is also important for scientists to determine the sequence of bases in a DNA molecule. Short segments of DNA can usually be sequenced in a few days; large segments may take longer. Scientists determine the order in which the new bases are added by synthesizing a new DNA strand on a template.

Many copies of a specific segment of DNA can be made through a process called a **polymerase chain reaction** (PCR). In certain cloning techniques, PCR is used to increase the amount of DNA. PCR has also been used to help diagnose human genetic disorders. In some criminal investigations, when only small amounts of DNA are available, PCR is used to increase the size of the sample for easy analysis. PCR results in millions of identical copies of a particular DNA sample.

DNA FINGERPRINTING

One aspect of biotechnology deals with DNA that is used to identify a person. Traditionally, identification has been made by identifying fingerprint patterns. Since no two people (except identical siblings) have exactly the same DNA sequence, it is possible to use these unique sequences as a means of identifying someone. This new technique has become known as **DNA fingerprinting.** This process is often used to compare a sample of DNA found in tissues collected at a crime scene with the DNA of the suspect.

Many people are concerned that certain applications of biotechnology will lead to possible abuses of individual rights. Decisions about the use of biotechnology often involve value judgments that will have to be decided by society. While we debate the uses and consequences of biotechnology, many benefits to society have already been achieved using these techniques, and new discoveries are being made daily.

QUESTIONS

Multiple Choice

1. A process that allows millions of copies of DNA to be produced is   *a.* DNA fingerprinting   *b.* PCR   *c.* gel electrophoresis   *d.* recombinant DNA.

2. Two sections of DNA that are joined together form   *a.* a ligase   *b.* a PCR   *c.* recombinant DNA   *d.* a clone.

3. Exact copies of DNA as a result of asexual reproduction are   *a.* clones   *b.* recombinant sequences   *c.* splices   *d.* complementary bases.

4. Circular pieces of DNA found in bacteria are called   *a.* clones   *b.* plasmids   *c.* ligases   *d.* endonucleases.

5. Chemicals used to cut DNA into fragments are   *a.* ligases   *b.* plasmids   *c.* clones   *d.* restriction enzymes.

Matching

6. genetic engineering
7. DNA fingerprinting
8. gel electrophoresis
9. PCR
10. DNA sequencing

a. determining order of bases in DNA

b. process to separate DNA fragments

c. making copies of DNA

d. manipulating the genes of living organisms

e. allows scientists to compare DNA sequences

Fill In

11. Gel electrophoresis works because DNA has a _____ charge.

12. Using gel electrophoresis, the size of a DNA fragment can be calculated by its rate of _____.

13. The process that takes a section of DNA from one organism and inserts it into the DNA of another organism is called gene _____.

14. _____ is the use of living organisms to help solve problems or make useful products.

15. The specific point where a restriction enzyme cuts a DNA sequence is called its _____.

Free Response

16. Explain the roles of restriction enzymes and ligases in the process of genetic engineering.

17. Explain the purpose of gel electrophoresis and tell how it works.

18. Why is PCR important in biotechnology?

19. Why are plasmids usually used to create recombinant DNA?

20. How can biotechnology improve life in the future?

UNIT SIX
CLASSIFICATION OF LIFE

Chapter 17

Taxonomy

ONE UNIQUE CHARACTERISTIC of humans is our need to organize our world. Most people find that organization makes every-thing—including biology—easier to learn. For example, your textbooks are organized into chapters to make learning quicker and easier. The food in the supermarket is organized under broad categories to make finding a particular item easier.

Biologists organize the world of living things in a classifica-tion system that groups similar living things together. Biologists use the word **taxonomy** to refer to their organization system. Taxonomy comes from Greek words that mean an "arrangement system."

EARLY ATTEMPTS AT CLASSIFYING ORGANISMS

The Greek philosopher **Aristotle** was one of the first per-sons to propose an organization system to classify living things. Aristotle lived more than 2000 years ago and had no micro-scope, computer, or textbooks to assist him in his work. Aris-totle grouped things based on the generalizations he made from his observations.

Aristotle classified all living things into either the plant or the animal group. Then he further grouped plants by size. All large plants were trees; low or medium-sized plants were shrubs; while the smallest plants were grouped with herbs.

Animals were further grouped by where they lived or how they moved. Aristotle grouped all flying animals together as air animals. Other animals were grouped as either land or water animals.

Although Aristotle made a very important scientific effort, his categories had major logical flaws and led to great difficulty in placing certain animals. For example, a frog lives both on land and in the water. With which group of animals would it be most logically placed? How could it be placed in one category but not the other?

If Aristotle had used more characteristics to group organisms he might have been more successful. For example, he might have noticed that aquatic bugs have much more in common with spiders (land animals) than with fish (water animals). Also, today we know that microscopic organisms exist. (These organisms do not fit in either the plant or animal kingdom.) Without modern technology, Aristotle had no way of knowing microscopic organisms existed at all.

MODERN CLASSIFICATION SYSTEMS

Through the years, the system of classifying life on Earth changed. Biologists wanted a system that would enable them to move away from the use of common names to describe organisms. It was confusing for a single organism to have several different common names, and often the common names were

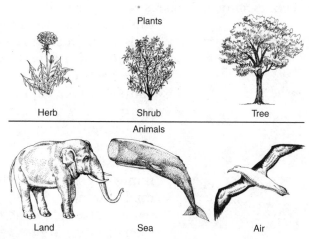

Figure 17-1. Aristotle's system of classification.

misleading. For example, a starfish is not a fish that is shaped like a star. It is, in fact, classified as an echinoderm, with other spiny-skinned animals.

The system of organization that is currently in use was actually first used, in a different form, in the seventeenth century. At that time, scientists began to classify organisms on the basis of physical characteristics. They did this to clear up some of the misconceptions that were formulated in Aristotle's system. They gave each organism a name called a **polynomial,** which literally means "many names." The first name in the polynomial was always the **genus** name. The genus that a plant or an animal belongs to contained a group of species with similar physical characteristics. The genus name was followed by a series of descriptive words. So the scientific name of an organism consisted of a series of words that described the organism in some detail. **Latin** was used as the universal scientific language so that scientists everywhere could discuss living things without confusion, even if they spoke different languages.

In the 1700s, a Swedish scientist named Carl von Linne (Carolus Linnaeus) invented a simpler, shorter system of naming living things with only two names. His system was called **binomial nomenclature.** Binomial nomenclature literally means "two-name naming." In this method, each organism received two names. As always, the first name was the genus name, the noun that named the group of species. This is always capitalized. For the second name, Linnaeus chose a specific adjective that would describe the group of animals or plants. This second descriptive word was known as the **species** name. The species name is never capitalized. Both words are written in italics or underlined. Scientists still use Linnaeus's system of naming organisms today. The **scientific name** of an organism is its binomial name. Some scientific names are *Homo sapiens* (humans), *Canis lupus* (wolf), and *Felis domesticus* (cat).

Figure 17-2. Carolus Linnaeus.

QUESTIONS

Multiple Choice

1. Aristotle classified plants by   *a.* where they lived *b.* their root structure   *c.* their stem height *d.* their flowers.

2. Long sentences used to describe organisms were called *a.* binomials   *b.* polynomials   *c.* uninomials *d.* definitions.

3. The language chosen to best suit classification systems was *a.* English   *b.* Spanish   *c.* French   *d.* Latin.

4. The scientific name is also the   *a.* polynomial   *b.* binomial *c.* uninomial   *d.* common name.

5. The first scientist to use binomial nomenclature was *a.* Linnaeus   *b.* Aristotle   *c.* Mendel   *d.* Sutton.

Matching

6. *Homo sapiens*   *a.* noun
7. species   *b.* wolf
8. genus   *c.* starfish
9. *Canis lupus*   *d.* human
10. common name   *e.* adjective

Fill In

11. The first word in a scientific name is the _____.
12. _____ classified animals by where they lived.
13. *Felis domesticus* is an example of _____ nomenclature.
14. In a scientific name, the _____ name is not capitalized.

Portfolio

See if you can develop a binomial system to name the members of your family. Use your family name as a genus name. Think up a species name that describes each member of your family.

Chapter *18*

Classification Systems

WHEN MODERN SCIENTISTS begin to classify and name a newly discovered organism, they rely on its appearance, or **morphology.** Taxonomists, the scientists who determine an organism's "place" in the natural world, examine an organism inside and out to determine its physical characteristics. Then they make comparisons to other known organisms. As they come closer to determining an organism's final classification and scientific name, taxonomists examine other aspects of its life that may not be so obvious. The taxonomist examines the organism through each phase of growth and development, observing its physiology and biochemistry. Today, taxonomists even study the organism's genetic information, to determine its relationship to other similar organisms. Finally, they are able to arrive at a name and place for the organism in the classification scheme of living things.

CLASSIFICATION SYSTEMS CHANGE OVER TIME

Aristotle classified organisms as either plant or animal. Even Linnaeus grouped all living things within these two kingdoms. Over time, with the aid of modern scientific equipment, scientists have added new kingdoms to the two original ones.

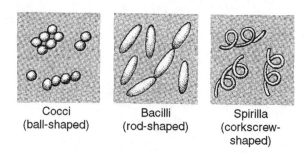

Cocci (ball-shaped)	Bacilli (rod-shaped)	Spirilla (corkscrew- shaped)

Figure 18-1. Monera.

Today, taxonomists place organisms in one of the following five major kingdoms: Monera, Protista, Fungi, Plantae, and Animalia. Any organism should have characteristics common to one of these kingdoms.

The **Kingdom Monera** is made up of one-celled organisms. They are prokaryotic, which means they do not have a membrane surrounding their nucleus. Monera is the only kingdom that contains prokaryotic organisms. Examples include bacteria and blue-green bacteria (formerly known as blue-green algae). Even though they are single cells, they are not simple organisms. They perform all the life functions that larger multicelled organisms perform, but they do it in a single cell. Monerans can be found everywhere—in the air, in water, and on land. They live in some of the most inhospitable environments on Earth. Some species even live in hot springs or in hot-water heaters! Monerans are present in living, nonliving, and dead material. Some make their own food by photosynthesis; others make their own food by using the energy trapped in the bonds of certain chemicals; still others obtain food by absorption. These organisms reproduce through **fission,** the process by which one organism divides into two separate organisms.

The **Kingdom Protista** also is made up of one-celled organisms. This kingdom was originally designed to contain organisms that could not easily be classified as a plant or an animal. Protists are eukaryotic, which means they have a membrane surrounding their nucleus and other organelles surrounded by membranes. Some examples include the ameba and the paramecium. Protists are commonly found in water or damp areas. Some make their own food through photosynthesis,

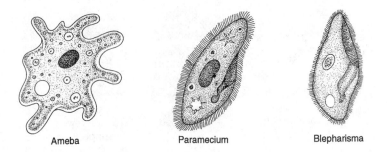

Ameba   Paramecium   Blepharisma

Figure 18-2. Protists.

while others ingest or absorb food from other sources. Protists reproduce sexually and asexually.

The **Kingdom Fungi** is made up of multicelled and some
one-celled organisms. Once grouped with plants, fungi now
have their own kingdom. Unlike plants, the cell walls of fungi do
not contain cellulose. Fungi also are not able to make their own
food. They must absorb nutrients from organic matter around

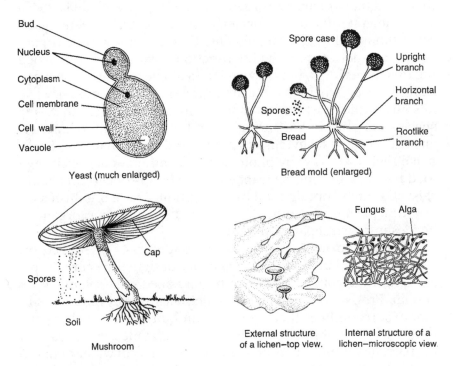

Figure 18-3. Fungi.

them. Fungi are eukaryotic. Common examples are mushrooms, yeasts, and molds. Some fungi cause diseases in other organisms. Athlete's foot is caused by a fungus, as are many diseases of plants. Most fungi live on land. Some species of fungi reproduce sexually, while others reproduce asexually.

Kingdom Plantae is one of the most easily identifiable and well-known kingdoms. All plants are multicellular and eukaryotic. Plants lack the ability to move from place to place. Most plants live on land. Almost all plants make their own food through photosynthesis. Most plants reproduce sexually. Some plants are able to reproduce asexually by runners, layering, or the use of cuttings. Some common plants are trees, flowering plants, shrubs, mosses, and ferns.

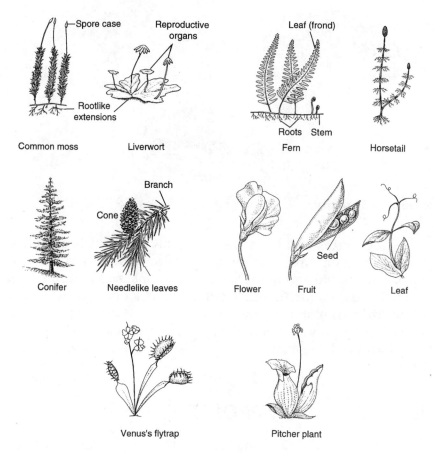

Figure 18-4. Plants.

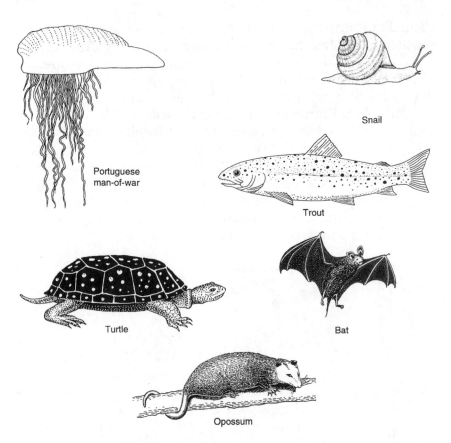

Figure 18-5. Animals.

Kingdom Animalia is another easily identifiable and well-known kingdom. Animals are multicelled and eukaryotic. Animals live on land and in the water. No animal can make its own food. Instead, animals must obtain food from their environment. Most animals reproduce sexually, but a few can reproduce asexually. Examples of organisms in the large and varied animal kingdom are kangaroos, worms, insects, sharks, sponges, and birds.

CLASSIFICATION PROBLEMS

Even with the five-kingdom system, some organisms are difficult to classify. Green algae, for example, can be single-celled or multicellular. They have been placed in the plant kingdom, even though plants are all supposed to be multicellular. Viruses present a classification problem to taxonomists. As technology enables us to discover new organisms and better identify organisms already classified, scientists will probably need to modify or replace the five-kingdom system.

Table 18–1. Characteristics of Organisms in the Five Kingdoms

	Kingdom				
	Monera	*Protista*	*Fungi*	*Plants*	*Animals*
Cells	Prokaryotes	Eukaryotes	Eukaryotes	Eukaryotes	Eukaryotes
Organization	Most are single cells	Most are single cells	Most are multicellular	Multicellular	Multicellular
Nutrition	Heterotrophs and Autotrophs	Heterotrophs and Autotrophs	Heterotrophs	Autotrophs	Heterotrophs

QUESTIONS

Multiple Choice

1. Aristotle's system of classification contained which two kingdoms? *a.* plant and fungi *b.* animal and protista *c.* plant and animal *d.* monera and protista

2. Most modern scientists use a classification system that contains *a.* two kingdoms *b.* three kingdoms *c.* four kingdoms *d.* five kingdoms.

3. Which of the following kingdoms contains mushrooms? *a.* Plantae *b.* Fungi *c.* Monera *d.* Protista

4. Which of the following kingdoms has members that do not have a membrane surrounding their nucleus? *a.* Monera *b.* Protista *c.* Plantae *d.* Fungi

5. The appearance of an organism is called its *a.* physiology *b.* morphology *c.* classification *d.* niche.

Matching

6. oak tree	*a.* fungi
7. dog	*b.* animal
8. mushroom	*c.* plant
9. paramecium	*d.* monera
10. bacteria	*e.* protista

Fill In

11. One-celled, prokaryotic organisms belong in the kingdom
    _____.

12. Members of the _____ kingdom are able to make their
    own food through the process of photosynthesis.

13. An organism that absorbs food from other organisms is in
    the _____ kingdom.

14. The monera and the _____ kingdoms both contain unicel-
    lular organisms.

15. Most monerans reproduce by _____.

Portfolio

Devise a classification system for a selection of 20 different
types of tools. Make sure that your system allows room for new
tools to be added when they are discovered.

UNIT SEVEN
THE MICROSCOPIC WORLD

Chapter *19*

Viruses

THE EFFECTS OF a viral disease, caused by the tobacco mosaic virus, were first studied about 100 years ago. Viruses themselves were not actually seen until the invention of the electron microscope. Most people are familiar with viruses through experiences with the common cold and other diseases caused by viruses. It is important to understand what viruses are, as well as how they affect living things.

WHAT IS A VIRUS?

The word *virus* itself means "poison." A **virus** is a noncellular particle that contains nucleic acid. A virus contains either DNA or RNA, never both, and is surrounded by a protein coat called a **capsid.** Because viruses have characteristics of both living and nonliving things, scientists are unable to classify viruses in any of the five major kingdoms of living things.

Viruses lack both cytoplasm and a cell membrane. They have no metabolism and never grow. Viruses can be crystallized and remain inactive for long periods of time. A living cell cannot be crystallized and retain the properties of living things. Viruses are inactive until they attach themselves to a cell. A virus can reproduce *only* when it is inside a living cell. The reproduction of

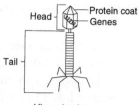

Head — Protein coat
— Genes

Tail —

Figure 19-1. Bacteriophage.   Virus structure

viruses is called **viral replication.** The cell a virus invades is called a **host.**

Viruses are classified by the type of cell they infect. Some viruses attack plant cells, some attack animal cells, and others invade bacterial cells. The type of virus that scientists know most about is the **bacteriophage**, which means "bacteria-eater." The name tells you that this group of viruses invades bacteria.

Viruses exist in several different shapes. The shape of the virus is determined by the arrangement of the protein molecules that make up its coat (capsid). Some viruses are shaped like rock crystals. The viruses that cause many respiratory infections have this shape. Another group of viruses are spiral shaped. The tobacco mosaic virus, which was studied in early viral experiments, is a spiral-shaped virus that attacks some plants.

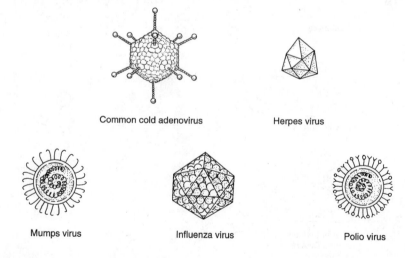

Common cold adenovirus   Herpes virus

Mumps virus   Influenza virus   Polio virus

Figure 19-2. Types of viruses.

VIRUSES AND DISEASE

A virus that causes a disease is termed **virulent.** If a virus stays inside its host for a long time without causing a disease, the virus is called **temperate.** The protein coating of the virus determines what kind of cell the virus can infect. A cell can be infected only if it has a receptor site for the protein of the virus. For example, a virus that might cause a sore throat can infect only cells in the throat, even though it might also come in contact with cells in the nose or cells on the hand. When the virus actually enters a cell, the result is an **infection.**

When viruses attack cells, they sometimes use the cells' proteins and genetic material to make new viruses. The host cell eventually breaks open, or lyses, releasing the new viruses. This process of making new viruses is called the **lytic cycle.** Some viruses go through a process called the **lysogenic cycle.** These viruses enter cells and become inactive until an external stimulus switches them on. The viral DNA actually becomes part of

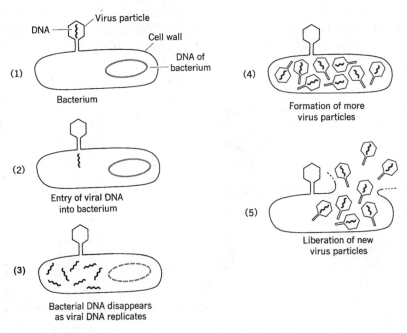

Figure 19-3. The lytic cycle.

the host cell's DNA. When the host cell makes copies of its own DNA, the viral DNA is included in those copies. This means that each time the host cell divides, the new recombinant DNA is passed on.

Viral diseases are very hard to control. Most antibiotics have no effect on diseases caused by viruses. Many common childhood diseases are caused by viruses. Measles, chicken pox, and mumps are all caused by viruses. There is strong evidence linking some viruses to certain cancers. At one time, viruses were thought to be the smallest carriers of disease. Now we know that some plant diseases are caused by small strands of RNA. These strands are called **viroids,** and are much smaller than the smallest known viruses.

THE AIDS VIRUS

One viral disease that is receiving a lot of attention today is **AIDS,** Acquired Immune Deficiency Syndrome. The AIDS virus weakens and sometimes completely destroys the immune system. This leaves the body unprotected and unable to fight infections. The virus is transmitted by the exchange of infected blood and other body fluids. Today, there is no cure for AIDS; however, treatments are available that affect the course of this disease. Scientists are working hard to find a cure for AIDS. It is best to avoid any possible infection with this particular virus.

QUESTIONS

Multiple Choice

1. The word *virus* means   *a*. bacteria-eater   *b*. poison
   c. small cell   *d*. parasite.
2. Viruses contain DNA or RNA and are wrapped in a coating made of   *a*. carbohydrates   *b*. lipids   *c*. nucleic acids
   d. proteins.
3. The viruses that invade bacteria are called   *a*. capsids
   b. bacteriophages   *c*. viroids   *d*. proviruses.

4. The cell a virus invades is called its   *a.* phage   *b.* host
   c. lytic cell   *d.* parasite.

5. When a virus invades and quickly destroys the cell, the
   virus is probably in the   *a.* lytic cycle   *b.* lysogenic cycle
   c. viroid cycle   *d.* temperate stage.

Matching

6. lytic cycle
7. lysogenic cycle
8. viroids
9. bacteriophage
10. virulent

a. virus that infects bacteria

b. viruses enter cell and become inactive for a period of time

c. virus that causes a disease

d. viruses enter a cell, take over, and reproduce rapidly

e. small strands of disease-causing RNA

Fill In

11. Viruses are not cells, they are _____.

12. The coating around a virus is a _____.

13. Viruses that enter cells but do not cause a disease immediately are called _____.

14. The tobacco mosaic virus is _____ shaped.

15. The reproduction of viruses is called viral _____.

Portfolio

Read *The Hot Zone* by Richard Preston. The book is a fascinating and chilling account of an outbreak of a deadly viral disease that appeared in Africa. Write a report about his book. You might like to form a group of classmates to discuss this book.

Chapter 20

Monerans

THE KINGDOM MONERA is made up of microscopic organisms commonly called **bacteria**. Advertisements constantly urge consumers to buy cleansers, mouthwashes, and disinfectants that will destroy the bacteria in their bodies and homes. From an early age, people are told to wash their hands before eating. While it is true that many species of bacteria are harmful and can cause sickness and disease, it is important to remember that other species of bacteria assist us in digestion, and food production, and even make chemicals we use as medicine. Monerans are a diverse group of organisms. If we learn more about their characteristics, we will better understand why bacteria have been so successful in surviving in many different environments on Earth.

MONERANS: THE PROKARYOTIC KINGDOM

The Kingdom Monera is the only kingdom made up of **prokaryotic** organisms. This means that monerans are organisms whose DNA is not contained within a nuclear membrane. In fact, bacteria lack any membrane-bound organelles such as mitochondria, vacuoles, or chloroplasts. The only membrane present in a moneran is its cell membrane.

Monerans are so small that we can see them only with a microscope. Most bacteria are only a few micrometers long. (A micrometer is one-millionth of a meter.) A moneran may live as a

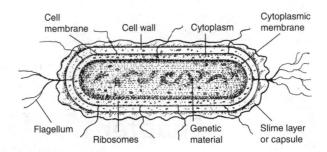

Figure 20-1. Bacterial cell.

single cell or it may live as part of a colony. A **colony** is a group of cells in which each cell functions independently.

Bacteria can live in any environment. Many of them possess flagella that are used for movement. Since bacteria lack a nuclear membrane, the DNA may appear as a single long chromosome, in a mass, or in small segments in the cytoplasm. Bacteria have a cell membrane and a cell wall. Some species of bacteria even have another outer protective layer called a capsule.

BLUE-GREEN BACTERIA AND TRUE BACTERIA

There are two major types of monerans: the blue-green bacteria (cyanobacteria) and the true bacteria (eubacteria).

Blue-green bacteria were formerly called blue-green algae. These organisms are **autotrophs,** or organisms that are able to make their own food. Blue-green bacteria make their food by photosynthesis, but they do not have chloroplasts. Instead, photosynthesis occurs on folds of the cell membrane. Blue-green bacteria have a cell wall, which gives the cell support.

Blue-green bacteria usually live in colonies and are found almost anywhere moisture is present. Many times, especially in polluted water, a population of blue-green bacteria undergoes a rapid increase in numbers. This is called a **bloom.**

The true bacteria, or **eubacteria,** usually feed on other organisms. Organisms that feed on other organisms are called **heterotrophs.** Some species of bacteria feed on dead organic matter. These bacteria are **saprophytes.** Other bacteria feed on living things. Organisms that feed in this manner are **parasites.**

REPRODUCTION AND GROWTH IN BACTERIA

Bacteria can increase their numbers very quickly. They can reproduce asexually or sexually. The asexual reproduction is called **fission.** This is a process by which a bacteria splits in half, forming two cells. Three types of sexual reproduction can occur in bacteria. However, sexual reproduction in bacteria is different from sexual reproduction in other organisms. In bacteria, sexual reproduction involves a sharing of DNA. This transfer of genetic material from one bacterium to another provides for the genetic diversity that makes bacterial species such good survivors. **Conjugation** is a method of sexual reproduction that involves the transfer of DNA from one living bacterial cell to another. **Transformation** occurs when a living bacterial cell absorbs the DNA from a dead bacterial cell. **Transduction** is the moving of bacterial DNA from one bacterial cell to another by the action of a virus.

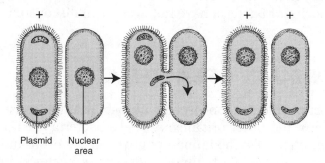

Figure 20-2. Conjugation in bacteria.

Several things are needed for bacteria to grow and multiply successfully. Certainly, bacteria need food and water. The cytoplasm of bacteria is about 80 percent water. One way to preserve food is by drying it. This causes bacteria present in the food to become inactive. Bacteria also need the proper temperature. Most bacteria grow best between 27° and 38° C. Cool temperatures slow bacterial growth, while freezing or boiling temperatures usually stop growth. Darkness is another condition that helps bacteria grow. Bright sunlight dries out bacteria, eventually killing them. The ultraviolet rays present in sunlight are also harmful to bacteria. Physicians use antibiotics to inhibit the growth of bacteria. Antibiotics have to be carefully designed to destroy harmful bacteria but not to eliminate useful bacteria.

CLASSIFICATION OF BACTERIA

The study of bacteria is called **bacteriology.** Bacteria are classified by their shape. Some bacteria are round and are called **cocci.** Other bacteria are rod-shaped and are called **bacilli.** Spiral-shaped bacteria are called **spirilla.** Prefixes can be added to the shape names to indicate the kinds of groups that are formed. *Diplo-* means "two." *Strepto-* indicates a long chain. A cluster of cells is shown by adding the prefix *staphylo-*. For example, streptococcus is a long chain of spherical cells. Strep throat is caused by streptococcus bacteria. A diplobacillus is two-rod shaped cells. Even though thousands of species of monerans have been identified, many more continue to be found. It is thought that most species of bacteria have not even been discovered yet.

IMPORTANCE OF BACTERIA

Most people are concerned with bacteria that cause disease in humans and other organisms. These bacteria are important, and their effects on the human population are profound. Today, medicines can deal with almost any type of bacterial infection, although people do still die from bacterial diseases. Because we mistakenly judge an organism's importance on Earth in relation to human needs, most of us are tempted to think that a world without bacteria would be a much better place to live. This idea couldn't be farther from the truth for many reasons. Without bacteria, it is true that certain human diseases would no longer occur. However, without blue-green bacteria, there would be a lack of oxygen. Dead organisms, which decompose as a result of the actions of bacteria, would no longer release beneficial substances back into the soil. Without bacteria, humans would lack certain vitamins. Bacteria present in our intestines produce important vitamins and help our bodies absorb them. Scientists certainly have more research to do in an effort to understand these prokaryotic cells that affect so many different organisms on Earth.

QUESTIONS

Multiple Choice

1. Organisms lacking a nuclear membrane are called
   a. heterotrophs *b.* autotrophs *c.* eukaryotic
   d. prokaryotic.
2. The cytoplasm in bacteria is about 80 percent *a.* DNA
   b. water *c.* chloroplasts *d.* mitochondria.
3. The moving of genetic material from one bacterium to another by a virus is *a.* transduction *b.* transformation
   c. fission *d.* conjugation.
4. Round bacteria are called *a.* bacilli *b.* spirilla
   c. diplobacillus *d.* cocci.
5. Bacteria can be classified by their *a.* speed *b.* size
   c. shape *d.* nucleus.

Matching

6. baccilli   *a.* group of bacterial cells
7. spirilli   *b.* spiral-shaped bacteria
8. colony   *c.* increase in population of blue-green bacteria
9. conjugation *d.* rod-shaped bacteria
10. bloom   *e.* sexual reproduction in bacteria

Fill In

11. Bacteria that feed on living things are called _____.
12. DNA in a moneran is located in the _____.
13. A cluster of round bacterial cells would be called a _____.
14. The blue-green bacteria are photosynthetic, but they do not have _____.
15. Bacteria that feed on dead organic matter are called _____.

Portfolio

The War of the Worlds by H. G. Wells is a brilliant work of science fiction about the invasion of Earth by aliens from outer space. The alien attack was stopped by an interesting twist of fate that relates to this chapter. Read this novel and report to the class on the aliens' defeat.

Chapter *21*

Protists

THE KINGDOM PROTISTA consists of a large group of microscopic organisms that live in moist environments. Protists are usually one-celled, but individuals of some species group together to form colonies. Because their cell has a nucleus surrounded by a nuclear membrane, protists are eukaryotes. Protists may be autotrophic or heterotrophic. **Autotrophic** organisms can make their own food; **heterotrophic** organisms depend on other organisms as a source of food. Some protists are like animals. They are called **protozoa.** Others are like plants. The plantlike protists are the **algae.**

THE PROTOZOA

The word *protozoa* means "first animal." Some protozoa are free-living and are found in lakes or oceans. Other protozoa are parasitic and survive only within other living organisms. Protozoa are all heterotrophic, and most are one-celled.

Most scientists classify protozoa based on how these organisms move. Protists in phylum **Sarcodina** move by forming projections with their membrane and cytoplasm. These projections are called **pseudopodia,** which means "false feet." Protists in phylum **Ciliophora** move by the use of tiny hairlike structures called **cilia.** The members of phylum **Zoomastigina** move by using tiny whips called **flagella.** The members of phylum **Sporozoa** have no means of locomotion. These protists are parasites and move along in the blood circulating in their hosts.

THE SARCODINES

There are several thousand species of sarcodines. Some sarcodines have an outer shell, others have an internal shell, and some have no shell at all. A common sarcodine that lacks a shell is the **ameba.** The ameba usually lives in freshwater. Unlike most organisms, the ameba has no definite outline. Its jellylike cytoplasm continually changes its shape. Amebas move by forming pseudopodia. The ameba eats a variety of foods, including other protists.

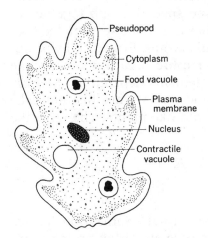

Figure 21-1. Ameba.

Most of the nutrients an ameba needs are absorbed by diffusion from the water it lives in, but some larger nutrients are not able to pass through the small openings in the ameba's cell membrane. These large molecules are taken in by the process of **endocytosis.** Two types of endocytosis have been identified: phagocytosis and pinocytosis. **Phagocytosis** engulfs large food particles and molecules. **Pinocytosis** takes liquids into the organism.

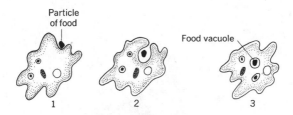

Figure 21-2. Phagocytosis.

Amebas also must take in oxygen for respiration and get rid of the carbon dioxide given off during respiration. The exchange of these gases occurs through the cell membrane.

Like all sarcodines, amebas reproduce by **binary fission.** Since this process is asexual, the offspring are identical copies of the first ameba.

It's amazing how these organisms can respond to the environment around them. Amebas respond to chemicals and light. If their environment becomes too harsh, amebas form **cysts,** hard protective outer walls that enable them to survive until the environment improves.

THE CILIOPHORES

The phylum Ciliophora is the largest group of protists. Organisms in this phylum live in freshwater and saltwater environments. The ciliophores are able to swim by using small hairlike structures called cilia. Cilia are made of protein fibers. A representative organism of this phylum is the **paramecium.** Unlike the ameba, the paramecium has a definite, unchanging shape. In fact, it looks like the outline of a shoe and is commonly called the "slipper" organism. It retains its shape because it is covered by a hard membrane called the **pellicle.**

A paramecium may have two nuclei. The **macronucleus** (large nucleus) controls most cellular processes. The **micronucleus** (small nucleus) controls reproduction. Reproduction can be either asexual or sexual.

Food enters the paramecium not through the membrane but through an **oral groove** leading to a mouth. Waste products

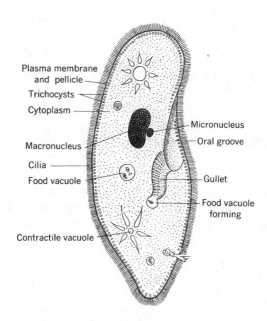

Figure 21-3. Paramecium.

pass out of an **anal pore**, located at the posterior, or back end. Paramecia also have a defense system made up of small poisonous darts called **trichocysts.**

THE ZOOMASTIGINA

The phylum Zoomastigina is composed of organisms that have one or more whiplike flagella. Flagella can push or pull organisms through water. Flagellates get their food from the freshwater or saltwater around them. Some members of this phylum are parasites. Trypanosomes, the organisms responsible for sleeping sickness, or trypanosomiasis, are flagellates.

THE SPOROZOANS

The sporozoans are parasitic organisms that lack structures designed for movement. Unlike other protozoans, sporozoans reproduce by producing spores. Each spore develops into an adult sporozoan. **Plasmodium** is an example of a sporozoan. Plasmodium is the parasitic organism that can cause the disease malaria in humans. It is transmitted to humans through the bite of the *Anopheles* mosquito. The plasmodium organisms are transferred to the human's blood and eventually reach the liver. In the liver, they reproduce and reenter the blood, causing red blood cells to break open. When a mosquito bites an infected person, the infected blood enters the mosquito, and the cycle continues.

THE ALGAE

Algae are plantlike protists. Some biologists still classify algae as plants, but most consider them protists due to their method of reproduction. Algae are usually classified based on color. Green algae are included in the phylum **Chlorophyta.** They seem to be the group of algae that modern plants are descended from. Some green algae are unicellular, while others are multicellular. **Spirogyra, Volvox,** and **Ulva** are all chlorophytes.

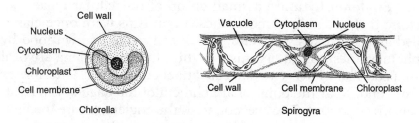

Figure 21-4. Green algae.

The red algae are included in phylum **Rhodophyta.** Red algae can be unicellular or multicellular. These algae are often found along rocky coastlines. A structure called a **holdfast** attaches them to the rocks. Many food products have substances in them that are derived from red algae, such as carrageenan and agar.

The brown algae are in phylum **Phaeophyta.** This is the group we normally call seaweed. Brown algae are larger than other kinds of algae. One common kind of brown algae, **kelp,** can grow to be more than 30 meters long. These algae have several unique structures that are adaptations to the environment of the sea. **Air bladders** help brown algae to float in the water. A stalk, called a **stipe,** gives the brown algae support. **Sargassum** and **Fucus** are other types of brown algae.

The **Chrysophyta** are the golden algae. These unicellular organisms are found in freshwater and saltwater. **Diatoms** are organisms in this phylum. Diatoms have shell-like walls made of silica—basically the same substance that makes up glass. The wall is made of two parts that fit together like the two halves of a petri dish. When diatoms die, their shells sink to the bottom of the ocean and form deposits of diatomaceous earth. Because of its gritty texture, diatomaceous earth is a gentle scouring agent. Diatomaceous earth is mined and used in toothpaste and scouring powders.

Fire algae, or **dinoflagellates,** are classified in phylum **Pyrrophyta.** Most fire algae are found in the ocean. The walls of dinoflagellates are made of cellulose. **Red tides** are caused by population explosions of pyrrophyta. Red tides are poisonous to shellfish and other animals in the ocean. Red tides also pose a danger to humans who consume shellfish contaminated with red tide organisms.

Euglenophyta are a small group of unicellular algae. Because they are able to move, these organisms were once classified as protozoans. **Euglena,** which can make its own food by photosynthesis, is an organism in this phylum. Euglena are oval shaped. They have two flagella attached at their anterior, or front end. Near the point of attachment of the flagella is a light-sensitive **eyespot.** The eyespot helps the euglena locate the light it uses for photosynthesis. Like most protists, the euglena reproduces by fission.

Even though the protists are very small, both algae and protozoans play important roles in maintaining the balance of

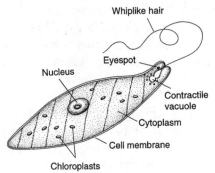

Figure 21-5. Euglena.

living things on Earth. They make up a vital part of Earth's food chains, and algae provide a great deal of Earth's atmospheric oxygen through the process of photosynthesis.

QUESTIONS

Multiple Choice

1. Animallike protists are called   *a.* algae   *b.* monerans   *c.* protozoa   *d.* protists.

2. The animallike protists are grouped according to their   *a.* color   *b.* method of movement   *c.* method of reproduction   *d.* habitat.

3. The hard membrane covering the outer membrane of a paramecium is the   *a.* micronucleus   *b.* pseudopod   *c.* pellicle   *d.* cilia.

4. Plasmodium is an example of a   *a.* ciliate   *b.* flagellate   *c.* sarcodine   *d.* sporozoan.

5. The structure in the euglena that helps it find light is the   *a.* eyespot   *b.* pellicle   *c.* macronucleus   *d.* flagella.

Matching

6. ameba   *a.* ciliophora

7. paramecium   *b.* sporozoan

8. Ulva   *c.* chrysophyta

9. diatoms   *d.* chlorophyte

10. plasmodium   *e.* sarcodine

Fill In

11. The phylum _____ makes up the largest group of protozoans.

12. A protist's _____ enables the organism to respond to light.

13. The brown algae are usually called _____.

14. Sporozoans are _____ organisms.

15. The false feet found on ameba are called _____.

Portfolio

Malaria is a very serious human disease with a fascinating life cycle. Find out more about malaria, a word that comes from the Italian words that mean "bad air." You might like to draw a diagram of the malaria life cycle or make a report for your class on this disease.

UNIT EIGHT
MANY-CELLED ORGANISMS

Chapter 22
Fungi

DID YOU EVER wonder what the black fuzzy growth on old bread, the green stuff that covers an orange, and the yeast that bakers use to make bread have in common? All are living organisms called fungi. The study of fungi is called **mycology.**

CHARACTERISTICS OF FUNGI

The Fungi Kingdom is made up of thousands of species of unicellular or multicellular organisms that obtain their nourishment from living and dead organisms through **absorption**. Absorption is the taking in of nutrients made by other organisms. Absorption is carried out by enzymes secreted from the fungi onto the food source.

Fungi are separated from plants because fungi feed by absorption, while plants are able to make their own food. Since fungi lack the chlorophyll used by plants (autotrophs) in the process of photosynthesis, they are heterotrophs. Some fungi are parasites, while others are saprophytes. **Parasites** feed on living organisms, and **saprophytes** feed on the organic material in dead organisms.

Most fungi reproduce by **spores,** which are carried through the air and water. A spore is a reproductive cell. Spores allow

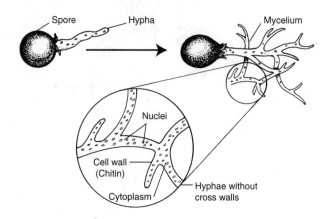

Figure 22-1. Hyphae.

fungi to travel and spread over large areas. Some fungus species reproduce by breaking apart, a process called **fragmentation.**

All fungi are composed of tiny threadlike structures called **hyphae.** Each hypha is covered with a cell wall made of **chitin.** Chitin is a hard carbohydrate material that protects the cells. As growth occurs, the hyphae branch and eventually form a mass of threads called a **mycelium.** This mycelium gives most fungi a fuzzy texture like cotton.

CLASSIFICATION OF FUNGI

Fungi are classified by their method of reproduction and by the shape of the structures formed by their hyphae. One group of fungi is called the **Zygomycota.** Most of these fungi live in the soil. Fungi in this group can reproduce asexually or sexually. *Rhizopus*, the common bread mold, is a member of this group. The hyphae of bread mold form a stalk. Spore cases, or **sporangia,** appear on top of the stalks. Spores are then released from the cases and carried by air currents to new areas where they germinate, or begin to grow.

Some zygomycetes actually live as part of the root system on a plant. A **mycorrhiza** is a symbiotic relationship between a fungus and a plant root. The fungus helps the plant's roots absorb the water and nutrients needed for its growth, while the plant's roots provide food from photosynthesis to the fungus.

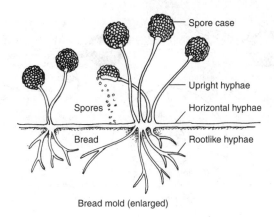

Bread mold (enlarged)

Figure 22-2. Bread mold.

Mushrooms, puffballs, and toadstools are classified in the group **Basidiomycota.** Members of this group are also called club fungi because of the club-shaped reproductive structures they form. These reproductive structures are called **basidia.** Basidia are located on the gills under the mushroom's cap. Gills are the thin structures that radiate out from the center of the mushroom like the spokes of a wheel. **Basidiospores** are produced in the basidia. Basidiospores are released into the air by the fungus. If they fall in a suitable place, they begin to grow and form new hyphae. Some members of this group are edible, while others are poisonous. You should never eat any mushrooms you find growing in nature, since they may be poisonous.

The sac fungi, or **Ascomycota,** make up the largest group of fungi. The members of this group have a saclike reproductive

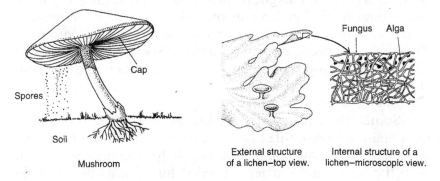

Figure 22-3. Mushroom and lichen structure.

structure called an **ascus. Ascospores** are released from the ascus and are carried through the air. If the spores land on a suitable surface, they germinate and form new hyphae. Yeasts, molds, mildews, and morels belong to this group. An important symbiotic relationship involves a member of this group. A **lichen** is a combination of an ascomycota fungus and an alga. In this symbiotic relationship, both the fungus and the alga are needed in order for the lichen they form to survive. The fungus provides a means of attachment and absorbs water from the environment. The alga provides food through photosynthesis. Together, the alga and the fungus form a very successful organism that is often the first living thing to colonize a new area.

Members of the group **Deuteromycota** cannot reproduce sexually. Therefore, these fungi are termed imperfect. The common blue and green molds seen growing on food are imperfect fungi. *Penicillium* is probably the best-known fungus in this group. The antibiotic penicillin is produced by this fungus.

IMPORTANCE OF FUNGI

The Fungi Kingdom is made up of a variety of organisms, some beneficial and some harmful. Many fungi, such as mushrooms, truffles, and yeasts, are edible or give tastes to other

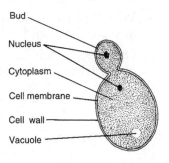

Figure 22-4. Yeast.   Yeast (much enlarged)

foods, such as certain cheeses. Yeasts are used in baking and in the manufacture of alcoholic beverages.

In nature, fungi play an important role as **decomposers.** They break down organic molecules into simple compounds.

The fungi use some of these compounds as food. The nutrients they don't use are made available for other organisms. Decomposers recycle many materials in the soil and provide raw materials for other processes.

Some fungi cause diseases in humans and other animals. Other types of fungi are important disease-producing organisms in plants. Athlete's foot and ringworm are two common human fungal diseases. Thousands of acres of crops are destroyed each year by parasitic plant fungi, such as rusts and smuts. However, some important medicines that are used to fight infections in humans are made from compounds that are derived from fungi.

QUESTIONS

Multiple Choice

1. The tiny threadlike structures that make up fungi are called   *a.* mycelium   *b.* hyphae   *c.* chitin   *d.* sporangia.

2. An organism that is a combination of an alga and a fungus is   *a.* chitin   *b.* ascus   *c.* mycelium   *d.* lichen.

3. Yeasts belong to the group   *a.* Basidiomycota   *b.* Ascomycota   *c.* Zygomycota   *d.* Deuteromycota.

4. Which of the following fungi have gills?   *a.* mushrooms   *b.* yeast   *c.* mold   *d.* mildew

5. Organisms that use dead organisms as a source of food are   *a.* parasites   *b.* autotrophs   *c.* saprophytes   *d.* mycorrhiza.

Matching

6. basidia   *a.* breaking apart of a fungus to reproduce
7. mycelium   *b.* carbohydrate material
8. fragmentation   *c.* mass of hyphae threads
9. chitin   *d.* reproductive structure
10. absorption   *e.* way of obtaining food

Fill In

11. _____ is the study of fungi.
12. The _____ contain the largest group of fungi.
13. Fungi are important _____ because they are able to break down organic compounds.
14. In a lichen, the _____ provides the food for the organism.
15. Fungi cells are protected by a covering of _____.

Portfolio

Imagine that you read in a newspaper article that a scientist has invented a spray that will kill all fungi in the world. To most people, it sounds like a good idea. However, you are a member of a blue-ribbon panel of ecology watchdogs, and your group opposes this idea for important biological reasons. Make a list of harmful changes that would occur on Earth if this spray was used. You might like to write a newspaper editorial detailing the reasons for your group's opposition to this project.

Chapter 23
Plants

THE HISTORY OF Earth is billions of years old—a long history marked by many changes. At some point in Earth's history, no plants were able to survive on land. Today, plants are widely distributed over Earth's surface. How did the tremendous greening of Earth happen?

Our present-day land plants probably evolved from algalike plants that lived in the oceans millions of years ago. Between 300 million and 600 million years ago, some of these algae became able to survive on land—at least for part of the time. Over time, two distinct groups of land plants with very different physical structures evolved. One group, the **bryophytes,** lack vascular tissue and have no true roots, stems, or leaves. **Vascular tissue** is made up of tubes that transport water and other materials throughout the plant. The second group, the **tracheophytes,** have vascular tissue. In addition, most tracheophyte species have true roots, stems, and leaves.

The plant kingdom is a group of multicellular organisms that are capable of making their own food. All plants are made of cells, and these cells form tissues, which form organs. The study of plants is called **botany.**

THE BRYOPHYTES: MOSSES, LIVERWORTS, AND HORNWORTS

Bryophytes are mostly small, nonvascular plants that usually grow in moist habitats. Bryophytes have no true roots, stems, or leaves. Since these plants have no vascular tissue, all

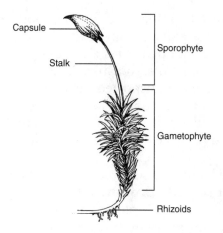

Figure 23-1. Moss.

the materials that move through them are transported by diffusion. This is a slow process and limits the size to which
bryophytes can grow. Instead of roots, **rhizoids** anchor
bryophytes in the ground. Bryophytes have stemlike and leaflike
structures, but they are certainly not as complex as the stems
and leaves found in the tracheophytes.

The moss life cycle will serve as an example of reproduction
in nonvascular plants. There are thousands of species of mosses.

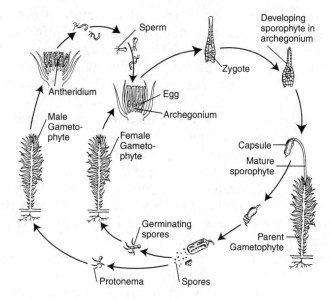

Figure 23-2. Life cycle of a moss.

Moss plants are either male or female. Both are small green plants, but the male plant has an **antheridium,** which produces sperm, and the female plant has an **archegonium,** which produces eggs. In these plants, moisture is needed for reproduction to occur. The sperm is transported through rain and other moisture on the plants to the female plant, where fertilization takes place. The fertilized egg develops into a stalk that supports a **capsule** at its end. Spores are produced inside the capsule. When the capsule breaks open, the spores are released and carried by the wind. If a spore lands in a suitable habitat, it begins to grow and forms a structure called a **protonema.** Each protonema develops into either a male or female plant, and the cycle continues.

Hornworts and liverworts are hardly noticeable. They resemble flattened leaves that lie on the ground—usually along the banks of streams or rivers, or in other damp places. Most of the time these plants reproduce sexually using antheridia and archegonia, but sometimes they reproduce asexually by producing small disks called **gemmae.** Each gemma can grow into a new plant.

TRACHEOPHYTES: THE VASCULAR PLANTS

The tracheophytes have three main organs—roots, stems, and leaves. The **roots** of a plant have several functions. The main purpose of the roots is to absorb and transport materials from the soil into the plant. Roots anchor plants in the ground, as you know if you have ever tried to pull dandelions out of a lawn. Roots also have the ability to store food.

The first root to grow is the **primary root.** If the primary root remains the largest root in the root system throughout the life of the plant, it is called a **taproot.** A carrot is an example of a plant with a taproot. In fact, the part of the carrot you eat is the taproot! Instead of one large root, some plants have a root system with many small roots. This is called a **fibrous** root system. Fibrous roots are excellent for holding soil in place. Lawn grasses have fibrous root systems.

Stems support a plant. In many plants, the stems hold the leaves so that they receive maximum exposure to light. Tubes in the vascular system transport materials through the stem to various parts of a plant. Plants also store food materials in stems.

Some plants have woody stems, and other plants have soft green stems. Stems can undergo primary and secondary growth. **Primary growth** is an increase in stem length. **Secondary growth** is an increase in the width or thickness of a stem.

Leaves are plant organs that are specialized for carrying out the process of photosynthesis. Their structure allows them to capture the energy in sunlight and to use this energy to convert raw materials—carbon dioxide and water—into sugars. Most leaves are composed of two parts, the blade and the petiole. The **blade** is the broad green part of a leaf. The **petiole** is the leafstalk that attaches the leaf to a stem.

Since tracheophytes have tubes to transport materials through their body, they are able to grow much taller than nonvascular plants. Two types of tubes are found in vascular plants: **xylem** carries water and other raw materials from the roots upward through the plant; **phloem** conducts food made in the leaves to other parts of the plant.

PLANTS THAT PRODUCE SEEDS

Some plants produce seeds, and others do not. Horsetails, club mosses, and ferns are vascular plants that reproduce by spores and not by seeds. These plants seem to be close relatives of plants that lived millions of years ago. Many of these primitive vascular plants are now extinct and are known only from their fossil remains.

Most plants produce seeds. There are two kinds of seed plants, gymnosperms and angiosperms. **Gymnosperms** are plants that produce unprotected or naked seeds. Gymnosperms can live to be very old. The bristlecone pine is one of the oldest living organisms in the world; it can live to be thousands of years old. **Angiosperms** produce seeds in a protective covering such as a nutshell or fruit. They also produce flowers. Some angiosperms are long-lived, while others live for only a few months.

Probably the best known gymnosperms are the conifers. **Conifers** are cone-bearing plants. Pine trees, redwoods, and firs are conifers. Each conifer produces male and female cones. The male cones usually grow in clusters and produce **pollen,** small grains that contain the sperm. The single female cones contain the ovules, which develop into seeds after they are fertilized. The cones also provide food for the developing seeds until they are released.

There are many thousands of different species of angiosperms. Angiosperms differ in their habit and timing of growth. Some are herbs. An **herb** is a nonwoody plant. Other angiosperms are shrubs or trees, plants that produce woody tissue.

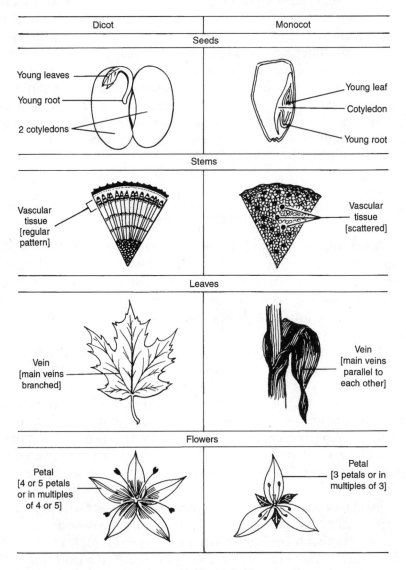

Figure 23-3. Features of monocots and dicots.

One reason for the success of angiosperms is that they have the ability to form **flowers,** which are very specialized reproductive organs. Seeds are produced in the flower. The flower eventually develops a **fruit,** which protects the seeds while they develop.

Scientists have divided angiosperms into two groups, the monocots and the dicots. **Monocots** have one seed leaf in their embryo. They usually have stems with scattered bundles of vascular tissue. Monocots also have fibrous roots and flower parts that are in three or multiples of three. Corn, wheat, oats, and rice are all monocots. **Dicots** contain two seed leaves in their embryo. They also possess taproots, flowers whose parts are in groups of four or five, and stems with vascular bundles in rings. Oaks, elms, magnolias, and maples are all dicots. In fact, most large plants with woody stems are dicots.

ALTERNATION OF GENERATIONS

All plants have a life cycle that includes an **alternation of generations.** This means that the plants produce two alternating generations of plants—a gametophyte stage and a sporophyte stage. A plant in the **gametophyte** stage produces gametes, sperm or eggs. These gametes combine to form zygotes, or fertilized eggs. Each zygote grows into a **sporophyte** plant. Sporophytes, or spore-producing plants, produce spores that develop into gametophyte plants, which are capable of producing more sex cells.

QUESTIONS

Multiple Choice

1. Plants that produce unprotected seeds are called
   a. angiosperms   *b.* ferns   *c.* flowers   *d.* gymnosperms.
2. All plants follow a life cycle called the   *a.* monocot cycle
   b. alternation of generations   *c.* vascular cycle
   d. protonema.
3. The study of plants is called   *a.* zoology   *b.* mycology
   c. cytology   *d.* botany.

4. In bryophytes, the organ in the male plant that produces sperm is the   *a.* archegonium   *b.* protonema   *c.* antheridium   *d.* phloem.

5. Which of the following is used to carry water and raw materials from the roots to the leaves?   *a.* xylem   *b.* phloem   *c.* rhizoid   *d.* gemma

Matching

6. nonvascular plants   *a.* a type of angiosperm
7. vascular plants   *b.* angiosperms
8. naked seed plants   *c.* gymnosperms
9. flowering plants .   *d.* bryophytes
10. monocot plant   *e.* tracheophytes

Fill In

11. An _____ is a nonwoody angiosperm plant.

12. Plants with _____ roots are excellent for holding soil in place.

13. In alternation of generations, the _____ stage produces spores.

14. Mosses and liverworts are examples of _____ plants.

15. _____ conducts food from the leaves to the rest of the plant.

Portfolio

Flowering plants supply most of our food supply. Find out why particular crops grow in certain areas. Rice, for example, grows in wet areas, while wheat grows in areas that receive moderate rainfall. What adaptations do these plants show that make them well-suited to different environments?

Chapter 24

Animals

PEOPLE COLLECT MANY things. Stamps and coins are popular items to collect. Libraries collect books and make the knowledge they contain available to people. Zoos are collections of animals. When you visit a zoo, you become aware of the great diversity of animals on Earth. Animals are found in many varied locations and in many varied forms. The study of animals, or **zoology,** can take a person from the depths of the ocean to the highest mountain and beyond! It's a journey of great excitement that you can share.

ANIMAL CHARACTERISTICS

Every animal has common characteristics that link it to other animals. All animals are eukaryotic, multicellular organisms. The cells of an animal lack cell walls. An animal is a heterotrophic organism that must take in food from an outside source to digest within its body. Most animals are capable of movement. Even animals that spend their adult life attached to one spot have a developmental stage when they are capable of movement. All animals reproduce sexually, and some can also reproduce asexually.

An important characteristic of most animals is body symmetry. Symmetry shows in an animal's body shape. Some animals show **radial symmetry**. In radial symmetry, the animal's body parts are in a circular arrangement around a central area, or axis. Sponges and jellyfish are animals that have radial symmetry. Most animals have **bilateral symmetry.** Animals with bilateral symmetry have a right and left side that are mirror images of each other. In animals with bilateral symmetry, it is also possible to identify the **anterior** (head or front end), **posterior** (tail or hind end), **ventral** (underside or belly), and **dorsal** (back) parts of the body.

More than 90 percent of the animal species on Earth do not have a backbone. These animals are **invertebrates.** The remaining animals that have a backbone are **vertebrates.** Because of the great diversity of animals, the animal kingdom is divided into many smaller groups called phyla.

PORIFERANS AND CNIDARIANS

The phylum **Porifera** includes the sponges. *Porifera* means "pore-bearing." All the animals in this phylum contain many small openings called **pores.** The pores allow water to circulate into and out of the body. Sponges rely on this movement of

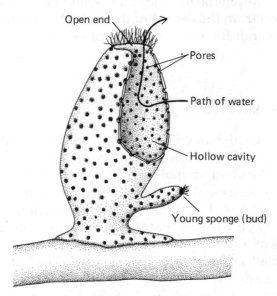

Open end
Pores
Path of water
Hollow cavity
Young sponge (bud)

Figure 24-1. Structure of a sponge.

water to get food and oxygen. Most members of this phylum live in the ocean. They also tend to remain in one location, not moving from place to place. Animals like the sponge that do not move are called **sessile** animals. Poriferans reproduce both sexually and asexually. Their asexual reproduction is called **budding.** Sponges can also grow back missing parts. This process is called **regeneration.** In sexual reproduction, a sperm fertilizes an egg, which develops into a larva. Sponge larvae are able to swim. Eventually the larvae settle and begin to develop into adult sponges.

Phylum **Cnidaria** contains jellyfish, hydras, coral, and sea anemones. Cnidarians have two different body forms. Some cnidarians show both forms at different stages of their life cycle. One form is a tubelike body, with tentacles that point up. Animals that are shaped like this are called **polyps.** A hydra is a

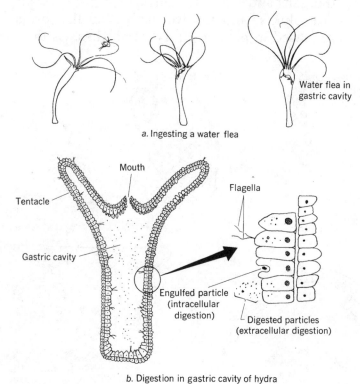

a. Ingesting a water flea

b. Digestion in gastric cavity of hydra

Figure 24-2. A hydra.

polyp-shaped cnidarian. Other cnidarians have bell-shaped bodies with tentacles that point down. These organisms are referred to as **medusae.** A jellyfish spends most of its life in this form.

One unique characteristic of this phylum is the **cnidoblast,** or stinging cell. Cnidoblasts are usually found on the tentacles and help the animal capture food. When an animal brushes against a tentacle, cnidoblasts are discharged, injecting venom into the animal. Two other features of cnidarians are a single digestive cavity and radial symmetry.

WORMS

Three phyla contain types of worms. Worms show bilateral symmetry and have three cell layers. They often have well-developed organs and organ systems.

Planaria, flukes, and tapeworms are all flatworms. Flatworms are placed in phylum **Platyhelminthes.** As their name

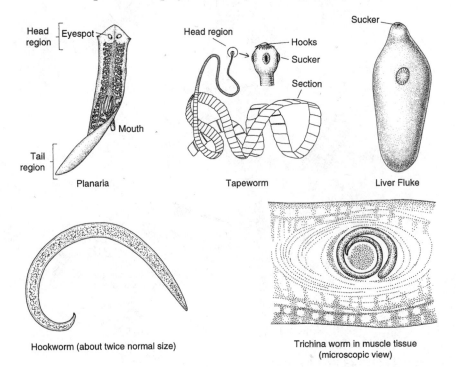

Figure 24-3. Some representative worms.

implies, these worms are flat. Flatworms have no body cavity and lack circulatory and respiratory systems. They have nerves and sense organs in their "head" region. Most flatworms are parasites that derive nutrition and protection from their host. Some, like the planaria, are free-living.

The phylum **Nematoda** includes the roundworms. Nematodes are very common yet rarely seen. These worms are usually very small, and they live in soil or water. Their body plan is a tube-within-a-tube, with a complete digestive system. There are a few parasites found in the phylum Nematoda. Roundworms, pinworms, hookworms, and trichinae are all nematode parasites that can infect humans.

Segmented worms are in phylum **Annelida.** Segmented worms are complex worms with well-developed systems. Earthworms and leeches are found in this phylum. Earthworms have no respiratory system. Oxygen is taken in, and carbon dioxide is given off, through the earthworm's skin. Each earthworm has both male and female reproductive organs. But a single earthworm cannot fertilize its own eggs. Earthworms are important organisms for improving the health of soil. Their wastes are used by plants. The tunnels they form as they move through the ground aerate the soil and improve plant growth. The annelids are the first animals to have a true coelom. A **coelom** is a fluid-filled body cavity.

MOLLUSKS AND ECHINODERMS

Another invertebrate group is the phylum **Mollusca**. Most mollusks live in the ocean, but some are found in freshwater and even on land. Mollusks are soft-bodied animals that usually have a shell or shells for protection. The largest mollusk class, **Gastropoda,** includes snails and slugs. These creatures slide along on a foot, eating a variety of plants, dead matter, or even living organisms. We recognize many members of the class **Pelecypoda** from the fish market. These are the bivalves, organisms that have two shells. These water-dwellers include clams, oysters, scallops, and mussels. The final group of mollusks are the **Cephalopoda.** Squid and octopuses are animals in this class. Both organisms are very efficient predators. Cephalopods are

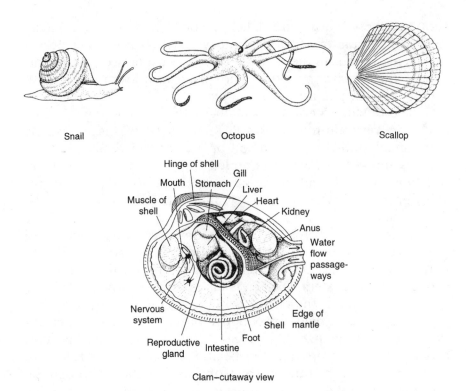

Snail   Octopus   Scallop

Clam–cutaway view

Figure 24-4. Mollusks.

the most highly developed invertebrates; their nervous systems and brains receive and process much information from the environment. Experiments have shown that octopuses can observe the actions of other octopuses and learn them. However, the ability of individual octopuses to learn varies greatly.

Many of the animals in phylum **Echinodermata** are known for their spines. *Echinoderm* literally means "spiny skin." This spiny skin is formed by a layer of spines and plates made of calcium found just below the skin's surface. All members of this class are invertebrates that live in saltwater. When echinoderms are larvae, they have bilateral symmetry. As adults, they have a type of radial symmetry, usually with five limbs. This is usually referred to as penta-radial symmetry. Animals in this phylum include sea stars, sea urchins, sand dollars, sea cucumbers, and brittle stars. Echinoderms reproduce sexually, and some are even able to regenerate, or regrow, lost body parts.

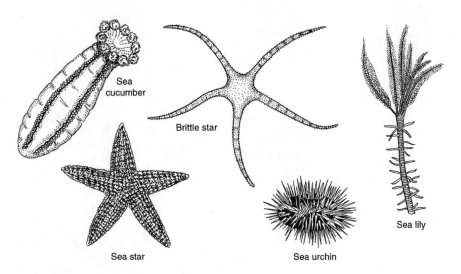

Figure 24-5. Echinoderms.

ARTHROPODS

The largest phylum of animals is the invertebrate phylum **Arthropoda**. *Arthropod* means "jointed leg." All arthropods have segmented bodies covered by an **exoskeleton.** The exoskeleton, located on the outside of the body, is made of **chitin.** Chitin is light in weight yet strong enough to protect the animal. There is one important disadvantage in having an exoskeleton. An exoskeleton must be shed before the animal can increase in size. This process of shedding the exoskeleton is called **molting.** During the time an arthropod is molting, it is more vulnerable to predation. Arthropods also have sense organs concentrated in their heads, which gives them a highly developed nervous system. Members of phylum Arthropoda are remarkably successful animals that live in a wide variety of habitats on Earth.

The arthropods are such a diverse group of animals that scientists have placed them into several subphyla and classes. The class **Crustacea** includes lobsters, crayfish, shrimp, crabs, barnacles, and water fleas. This group of animals serves as an important food source for other animals. Most crustaceans live in the ocean. They are unique in the variety of appendages they have developed for movement and feeding. Various species of crustaceans have either eight or ten legs.

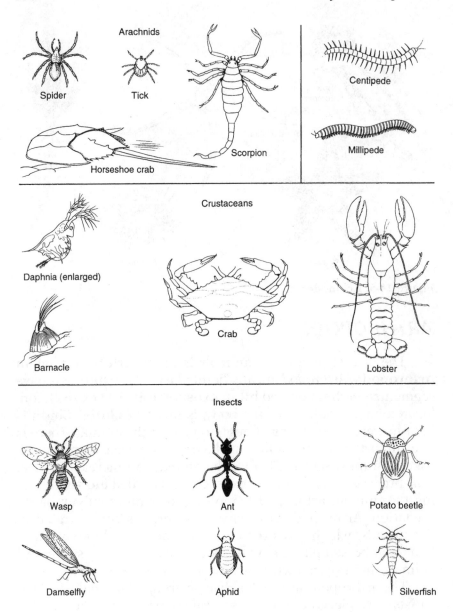

Figure 24-6. Arthropods.

Spiders, scorpions, and ticks belong to the subphylum **Arachnida.** All arachnids have eight legs and two body parts. One unique structure found in this subphylum is the **spinneret.** The spinneret releases silk from the body of arachnids. The silk produced by these animals is stronger than steel wire of the same size. This silk is used to make webs and other structures. Another unusual adaptation in this group of animals is the structure called **book lungs.** Book lungs allow for gas exchange in arachnids.

The subphylum **Insecta** makes up the largest group of arthropods. The study of insects is called **entomology.** Insects are the only invertebrates capable of flying. Insects have three body regions and six legs. The head region has antennae, simple eyes, and mouthparts. The mouthparts are adapted to various methods of eating.

Insects go through several changes before they reach the adult stage. Some insects follow a series of developmental changes called **incomplete metamorphosis.** Animals that undergo incomplete metamorphosis pass through the stages of egg, nymph, and then adult. A grasshopper is an insect that undergoes incomplete metamorphosis in its development.

Complete metamorphosis has four distinct stages: egg, larva, pupa, and adult. Butterflies undergo complete metamorphosis.

Some insects are harmful to people, but many species are helpful. They pollinate plants and serve as an excellent food source for other animals. Insects also seem to have very complex behavior patterns. Social insects, such as ants and bees, cooperate to benefit their whole colony. This cooperation is called **division of labor** and is very important for the survival of social insects.

Myriapods are many-legged arthropods. The classes **Chilopoda** and **Diplopoda** contain different types of many-legged arthropods. Centipedes are members of the Chilopoda class, and millipedes are members of the Diplopoda class. Centipedes have one pair of legs per segment. Millipedes have two pairs of legs for each segment. Centipedes eat other animals, while millipedes eat plants.

THE CHORDATES

All other animals are placed in phylum **Chordata.** Members of this phylum have several characteristics that distinguish them from animals in the other phyla. All chordates have a **no-tochord.** A notochord is a stiff, rodlike structure that usually runs down the back of the animal. Chordates have a **dorsal nerve cord** that carries information to the brain. **Gill slits** are also found in chordates. In many species, gill slits disappear before the animal completes development.

The phylum Chordata is divided into three subphyla. The subphylum **Urochordata** contains sessile marine animals. Sea squirts belong in this group. Lancelets *(Amphioxus)* belong to the subphylum **Cephalochordata.** Lancelets usually live along the seacoast, partially buried in the sand. They keep their heads out of the sand, so they are able to filter the water for food. The best-known subphylum of chordates are the **Vertebrata.** Members of this subphylum are distributed widely throughout the

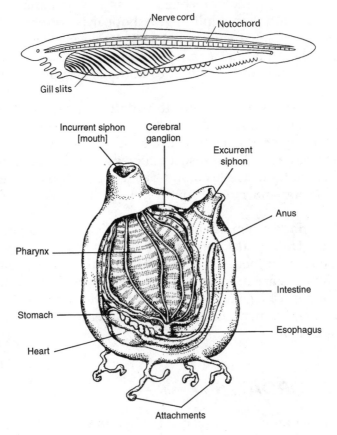

Figure 24-7. Lancelet and sea squirt.

world. Vertebrates have internal skeletons, including bones that surround and protect their nerve cord. These bones are called vertebrae and make up what is commonly known as our back-bone. Fish, amphibians, reptiles, birds, and mammals are all classes of animals included in the subphylum Vertebrata.

There are three classes of fish. The class **Agnatha** contains the jawless fish. The lamprey and the hagfish are included in this class. They are eellike parasitic fish that attach themselves to other fish and feed on blood and body tissues.

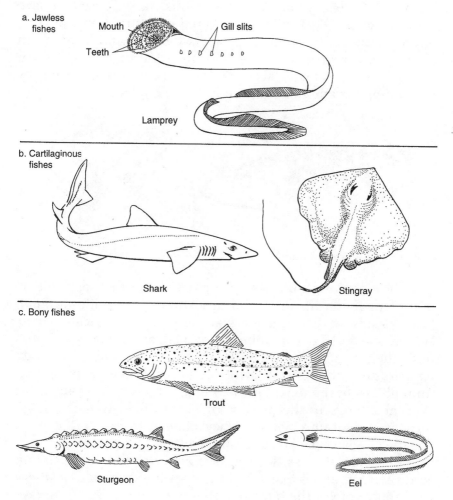

Figure 24-8. Fish.

Fish in the class **Chondrichthyes** have jaws and skeletons that are made of cartilage. Sharks, skates, and rays are included in this class. Many members of this class have a rough skin covered with **placoid scales,** small, toothlike projections. These scales make the skin so rough that at one time it was used as a kind of sandpaper. These fish lack a covering over their gills. If you have ever observed a shark, you may have seen slits in the head area. After water passes over the gills, it moves out of the shark's body through these slits.

The majority of fish belong to the class **Osteichthyes.** Any fish you have eaten most likely is in this class. These fish have bony skeletons. Their gills are covered by a structure called an **operculum.** The operculum protects the gills and keeps them clean. Another structure found in bony fish is the swim bladder, a gas-filled sac. A fish is able to adjust its depth in the water by controlling the amount of gas in its swim bladder.

Salamander   Frog

Figure 24-9. Amphibians.

The class **Amphibia** is made up of animals that spend a portion of their life in water and a portion of their life on land. Frogs, toads, and salamanders are amphibians. Because their eggs lack a protective membrane or shell, amphibians must deposit their eggs in the water. Like fish, amphibians have body temperatures that change with the surrounding environment. Amphibians have moist, thin skin with no scales. They use gills, skin, and lungs in the process of respiration. Amphibians go through a **metamorphosis.** They change from an immature aquatic (water) stage to an adult terrestrial (land) stage.

Members of class **Reptilia** are better adapted to a life on land than are amphibians. Reptile eggs have a tough, leathery shell that protects them from drying out on land. Reptiles have a dry skin that is covered by scales. Some reptiles—snakes and

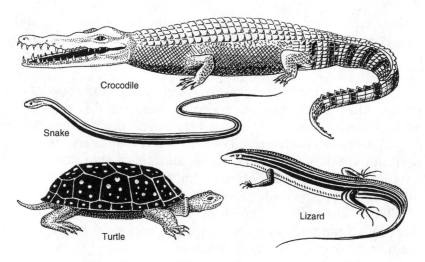

Figure 24-10. Reptiles.

certain lizards—have no legs and feet. Others have feet with claws or pads for better movement on land. Members of the class Reptilia include snakes, lizards, turtles, alligators, and crocodiles.

Birds are placed in the class **Aves**, composed of nearly 9000 species. Birds are able to regulate their body temperature internally. As a result, birds can live in a variety of habitats that range from the south pole to the tropics. Characteristics of birds

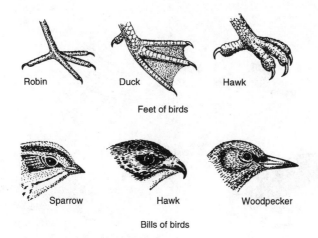

Figure 24-11. Characteristics of some birds.

include wings, hollow bones, beaks, shelled eggs, feathers, and a four-chambered heart. Birds can be classified by their beaks, feet, or the way they fly. There are even several kinds of flightless birds.

Most people are familiar with animals in the class **Mammalia.** After all, humans are also included in this class. Like birds, mammals are able to maintain their body temperature. Mammals also have a four-chambered heart. Body hair is an important mammalian characteristic. Mammary glands are found only in mammals. Nourishment for the young in the form of milk is provided by these glands.

There are only two types of mammals that lay eggs—the duck-billed platypus and the spiny anteater, or echidna. These mammals are called **monotremes.**

Marsupials are pouched mammals. Young marsupials emerge at a very early stage in their development. They hold onto their mother's fur as they crawl into her pouch for further development. The only marsupial in North America is the opossum. Kangaroos and koalas are also marsupials. These animals (and most other marsupials) are native to Australia.

The most successful group of mammals today are the **placental** mammals. Humans are placental mammals. These mammals have a special structure called a placenta, which allows the young to develop within the mother's body for a period of time.

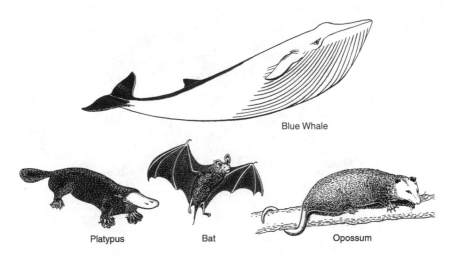

Blue Whale

Platypus   Bat   Opossum

Figure 24-12. Mammals.

An umbilical cord connects the unborn animal to the placenta. This cord allows for the transfer of nourishment from the mother to the fetus. Wastes produced by the fetus pass through the umbilical cord to the mother. Your belly button is the physical remains of this attachment to your mother.

By observing similarities and differences in body structure, we can group animals into phyla. Through the study of similar animals, we begin to understand the importance of the relationships among all animals and the survival of the Kingdom Animalia on Earth.

QUESTIONS

Multiple Choice

1. Animals that are characterized by pores are the
   a. mollusks   *b*. sponges   *c*. vertebrates   *d*. jellyfish.

2. The study of insects is called   *a*. zoology   *b*. botany
   c. ichthyology   *d*. entomology.

3. Fish, amphibians, reptiles, birds, and mammals are in
   the phylum   *a*. Vertebrata   *b*. Animalia   *c*. Chordata
   d. Cephalochordata.

4. Sharks, which have cartilage skeletons, are members of
   the class   *a*. Agnatha   *b*. Chordata   *c*. Osteichthyes
   d. Chondrichthyes.

5. The study of all animals is called   *a*. zoology   *b*. botany
   c. virology   *d*. physiology.

Matching

6. Amphibia   *a*. dog

7. Reptilia   *b*. perch

8. Aves   *c*. turtle

9. Mammalia   *d*. frog

10. Osteichthyes   *e*. robin

Fill In

11. An animal's body plan that lets you distinguish between a right side and a left side is called _____ symmetry.
12. The stinging cells found on jellyfish are called _____.
13. A _____ is a fluid-filled space found in an animal.
14. The largest phylum of animals is the group called _____.
15. Humans are classified as _____ mammals.

Portfolio

Dinosaurs were once common on Earth, but they are now extinct. Research a specific kind of dinosaur. Find out what kinds of foods it ate and where it lived. Make a model or a drawing of the dinosaur you are researching. Share your information with your class.

UNIT NINE
BIOLOGY OF HUMANS

Chapter 25
Human History

A WALK DOWN a street in a modern city clearly shows the effects of modern technologies on people's lives. Cars, buses, electric lights, the availability of information in both written and visual forms are everywhere. But even from your own experiences you know that life was not always as it is today. The long history of Earth is marked by constant change.

Over time, humans have also changed. As you study biology, your natural curiosity about the development of the human race may be enhanced. You may even begin to wonder about our early ancestors and how we are related to other animals in the world.

THE HUMAN PAST

Anthropologists are scientists who study humans and attempt to answer these kinds of questions. They study civilization and different cultures throughout the world and make comparisons between the ways different people live. They also study the human past in ways that are different from the ways historians study our past. Anthropologists search through fossils to find out how prehistoric humans lived. These studies show how we have changed to become the people we are today, both in body and habit.

Anthropologists have named the earliest ancestors of humans hominids. **Hominids** are humanlike mammals that walked upright on their hind legs. From examining fossil teeth, scientists have determined that hominids were mostly omnivores that ate both vegetables and meats. In these ways, early hominids were much like modern humans.

Many fossils of these early hominids have been found in Africa, where humans evolved. In 1924, Professor Raymond Dart named one group of these fossils ***Australopithecus africanus,*** which means "southern ape from Africa." This name is appropriate because although these fossils may be remains of our early ancestors, their brain size is more like that of an ape. Brain size is only one of the many indicators of animal relationships that scientists use. A scientist determines the brain size of fossil hominids, modern apes, and people by examining the size of their skull. The larger the skull, the larger the brain it enclosed and protected.

THE WORK OF THE LEAKEY FAMILY

Three anthropologists who have identified several important pieces of the puzzle that makes up the human past are Mary and Louis Leakey and their son Richard. When they dug up a hominid skull with a much larger braincase than that of *Australopithecus*, they knew they had found a new species of human. They named this fossil ***Homo habilis.*** Through scientific-dating techniques, it was determined that *Homo habilis* lived much later than *Australopithecus africanus.* Homo habilis means "handy human" and was so named because it used crude stone tools. Some anthropologists feel that some of these tools could have been used to kill animals. Tool marks found on animal bones indicate that *Homo habilis* may have hunted and eaten meat. However, other anthropologists feel that *Homo habilis* was mainly a vegetarian that ate meat when it was available.

Scientists have found a hominid with an even larger brain. This species is known as ***Homo erectus*** (named for its upright posture) and has more characteristics in common with modern humans. *Homo erectus* lived in groups that migrated from Africa through Europe and throughout Asia. Similar human fossils

Figure 25-1. Homo erectus.

have been found in all these places, along with more advanced stone tools. Scientists have even found evidence that *Homo erectus* used fire.

The most modern of the hominids had braincases equal in size to our own. The scientific name given to a hominid with this large-sized brain is ***Homo sapiens.*** The earliest fossils of *Homo sapiens* were found in Europe. The **Neanderthals** were a species that had large, thick skeletons. They hunted with advanced stone tools and lived in shelters in groups. They may even have had some religious beliefs because they buried their dead. The Neanderthals eventually became extinct.

Another species that lived at the same time as the Neanderthals were the **Cro-Magnons.** Fossils of these hominids have been found in the area that is now France. Cro-Magnons are known today for their magnificent cave art. The Cro-Magnons had skeletons that are the most like those of modern humans. They were probably taller than the Neanderthals. Cro-Magnons

Figure 25-2. Neanderthal.

lived in groups and used a variety of complex tools made from stone and other natural materials. They hunted for their food, made clothes, used fire, lived in shelters, and are considered to be direct ancestors of modern humans.

PRIMATES

As scientists have worked to trace early human history and to classify new species, our relationship to other animals has become clearer. Humans are **primates,** mammals with large brains that live in groups with complex social interactions. Other primates include apes, monkeys, and lemurs. Like most primates, our earliest ancestors must have lived in trees. This is made evident by examining some of the adaptations that are common to all primates.

An **adaptation** is a genetic trait that helps an organism survive in its environment. To survive, early primates had to be safe from predators and have a source of food. They accomplished both needs by living in the trees, away from the dangers posed by ground-living predators. Swinging from branch to branch required a shoulder structure that would allow free range of movement. To better grasp branch tips, primates evolved an opposable thumb. Having the ability to touch the thumb to other fingers and parts of the hand also enabled the primates to grasp food and eventually use tools. Finally, primates evolved forward-facing eyes. Forward-facing eyes allow for a stereoscopic, or three-dimensional, field of vision. This kind of vision provides animals with better depth perception, an enhanced ability to judge distance and dimension. Greater depth perception provides animals with an ability to move more safely from branch to branch.

Eventually, our human ancestors left the trees to walk on the ground. Their large brains enabled them to develop the advanced skills that permitted them to adapt to new environments. Apes, monkeys, and lemurs still cling to the trees for safety from predators, which unfortunately now also include their human primate relatives.

QUESTIONS

Multiple Choice

1. Humanlike ancestors that walked on hind legs are
   a. *Homo sapiens*   *b.* hominids   *c.* apes   *d.* lemurs.

2. Scientists who try to explain human origins are called
   a. anthropologists   *b.* botanists   *c.* zoologists
   d. geneticists.

3. Humans, apes, monkeys, and lemurs are called
   a. *Homo sapiens* *b.* Cro-Magnons   *c.* primates
   d. Neanderthals.

4. The most recent ancestor of modern humans is
   a. Neanderthal   *b.* *Homo habilis*   *c.* Australopithecus
   d. Cro-Magnon.

5. Which of the following ancestors of modern humans were responsible for the cave art found in France?
   a. Cro-Magnons   *b.* Neanderthals   *c. Homo sapiens*
   d. Homo habilis

Matching

6. fossils
7. *Australopithecus africanus*
8. *Homo sapiens*
9. Neanderthals
10. *Homo erectus*

a. modern humans
b. found only in Africa
c. remains of early humans
d. early *Homo sapiens*
e. migrated through Europe

Fill In

11. An _____ is a trait that helps an animal survive in its environment.
12. The _____ thumb allows primates to grasp objects.
13. *Australopithecus africanus* was discovered in Africa in _____.
14. _____ is the name given by anthropologists for the earliest ancestors of humans.
15. The earliest examples of *Homo sapiens* were found in _____.

Portfolio

Imagine that you are in charge of making a time capsule that will provide future generations of anthropologists with evidence of the kind of life you lived toward the end of the twentieth century. What kinds of items would you include? Why did you make your selections?

Chapter 26
Skeletal and Muscular Systems

ATTEND A SCHOOL dance and you will observe only some of the movements the human body is capable of making. Twirls, dips, slides, glides, twists, and wiggles all show ways the body is capable of moving. How are these movements possible? The human body is able to move because of the interaction between the skeletal and muscular systems.

THE SKELETON

The body of an adult human has 206 bones and more than 600 muscles. The bones provide a frame for our body called the **skeleton.** The connecting ligaments, tendons, and muscles allow us to move the bones that make up our skeleton. Individual bones and muscles perform a variety of specific jobs to keep our body moving, safe, and alive.

Our mammalian skeleton resembles the skeleton in fish, reptiles, amphibians, and birds in one important way. Unlike the exoskeleton that surrounds an insect or a lobster, an **endoskeleton** is found inside a protective covering of skin and muscles. Our skeleton, cushioned from breaks by layers of skin and muscles, also has the job of protecting many of our vital organs. For example, the cranium, or skull, protects the brain; the spinal column protects the spinal cord; and the rib cage protects our heart

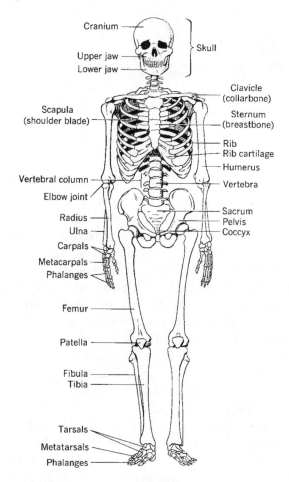

Cranium

Upper jaw

Lower jaw

Skull

Clavicle (collarbone)

Scapula (shoulder blade)

Sternum (breastbone)

Rib

Rib cartilage

Humerus

Vertebral column

Vertebra

Elbow joint

Radius

Sacrum

Pelvis

Ulna

Coccyx

Carpals

Metacarpals

Phalanges

Femur

Patella

Fibula

Tibia

Tarsals

Metatarsals

Phalanges

Figure 26-1. Human skeleton.

and lungs. The skeleton also supports our internal organs, and with our muscles gives our body its shape and appearance. Minerals are stored in our bones, providing a reserve that can be used when needed by tissues throughout the body.

The human skeleton is divided into two main divisions. The **axial skeleton** is made up of all the bones that lie down the center of the body and protect the internal organs. The skull, vertebrae, and thorax make up the axial skeleton. The skull protects the delicate tissue of our brain. The spinal cord, the main route for carrying impulses through our nervous system, is protected by the vertebrae. The thorax, composed of the sternum and

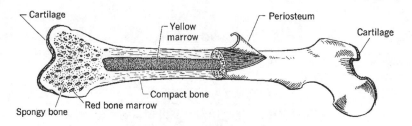

Figure 26-2. The femur or thigh bone.

the ribs, protects the heart, lungs, liver, and spleen. Bones outside the axial skeleton are part of the **appendicular skeleton.** The shoulder, arm, hip, and leg bones are part of the appendicular skeleton. These bones usually play a role in body movements.

The long bones in the legs and arms, the pelvis, ribs, sternum, and vertebrae contain **red bone marrow.** This marrow makes red and white blood cells. The long bones in the arms and legs also contain **yellow bone marrow.** This marrow is made up of fat cells and is actually a form of stored energy.

Each bone in our body is developed through a process called **ossification.** One type of ossification occurs when layers of membrane form and later harden. Some bones that develop into flat, platelike shapes ossify in this way. Examples of this type of bone are the plates of the skull and the collarbone.

Ossification can also occur through the changing of cartilage to bone. **Cartilage** is a type of flexible connective tissue. Cartilage makes up the entire skeleton of sharks. You can also find cartilage in the places where the different parts of a chicken leg meet. Cartilage is sometimes called gristle. During the first month of pregnancy, the skeleton of a human embryo is made entirely of cartilage. Gradually, minerals are released into the cartilage as the fetus develops. These minerals cause the cartilage to harden into bone. This process continues through birth and into the teenage years until most of our skeleton has become bone. Adult humans do retain some cartilage in joints like the knee, in our nose, and in our ears.

The place where two bones meet is called a **joint.** There are many different kinds of joints. Each type of joint moves in a special way, allowing the body to move its parts in various directions. Some joints, like the fixed joints in the skull and the semimovable joints in the spine, move very little if at all. These joints are relatively strong and protect and support various body organs. The pivot, hinge, angular, ball-and-socket, and gliding joints are movable joints that allow the body to show flexibility and a wide range of different motions. If the bones rubbed together at the joints, movement would be painful. The body produces a special fluid, called synovial fluid, to aid movement at the joints. **Synovial fluid** protects the ends of the bones from the constant pressure, friction, and wear that occur when the bones move. The bones at a joint are held together by tough bands of connective tissue called **ligaments**. Ligaments connect one bone to another and are flexible enough to stretch as the bones move.

THE MUSCLES OF THE BODY

Joints are the bones' connection sites and are extremely important for body movement. However, our bones are just a framework and could not move without the connections to our skeletal muscles. For example, when you move your leg, some skeletal muscles contract while others relax. This contraction and relaxation of muscles moves the bones in your leg.

Skeletal muscles are connected to bones by a special form of connective tissue called **tendons**. Skeletal muscles are made of individual fibers. Skeletal muscles are also called striated

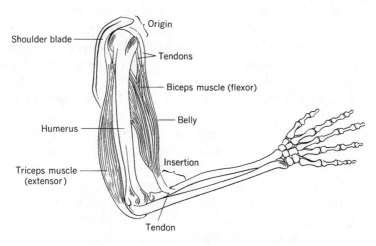

Figure 26-3. Bones and certain muscles of the arm.

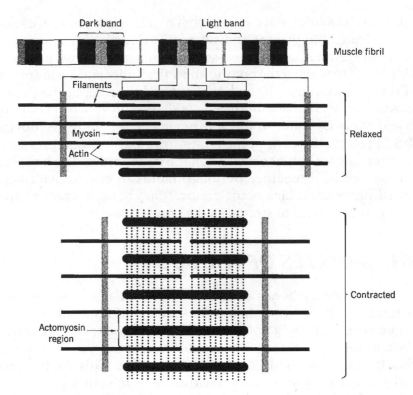

Figure 26-4. Muscle fibers.

muscles because the muscle fibers have light and dark cross-
bands. These are voluntary muscles because we can choose
when and how to move them. We can go to a gym and exercise
to improve their performance, strength, shape, and appearance.

A person's large muscles are made up of thousands of fibers.
These fibers are made of smaller protein units called **myofibrils.**
Each myofibril is made up of units called **sarcomeres.** Sarcomeres consist of two proteins: actin and myosin. **Actin** is found
in very thin filaments, and **myosin** is found in thick filaments.

Muscles contract when they are stimulated by a nerve impulse. Muscles become shorter when they contract. The protein
filaments that make up the muscle fibers actually slide past each
other. If the filaments slide toward each other, the muscle shortens or contracts. If the filaments slide away from each other, the
muscle relaxes and returns to its normal length. This movement
is called the **sliding filament theory** of muscle contraction.

There are two other types of muscles found in the human body: cardiac and smooth muscle. The heart is made of striated muscle called **cardiac muscle.** Heart muscle is striated but not as clearly as skeletal muscle. Cardiac muscle is an involuntary muscle because we cannot consciously control its contractions. All of the muscle fibers in our heart contract at the same time, and all of them relax at the same time. Cardiac muscle cells beat together because of the sinoatrial node, which sends electrical nerve impulses throughout the heart. **Smooth muscle** also is an involuntary muscle. It is located in our digestive system, bladder, and the walls of the blood vessels. Smooth muscle is not striated and contracts more slowly than does skeletal muscle.

QUESTIONS

Multiple Choice

1. Muscles are connected to bones by   *a.* ligaments
   b. tendons   *c.* synovial fluid   *d.* cartilage.
2. Muscle fibers are made of smaller units called   *a.* joints
   b. smooth fibers   *c.* myofibrils   *d.* ligaments.
3. The skull and thorax are part of the   *a.* appendicular skeleton   *b.* exoskeleton   *c.* axial skeleton   *d.* cartilage system.
4. The place where two bones meet is called a   *a.* joint
   b. sarcomere   *c.* tendon   *d.* ligament.
5. The lining of the intestines is made of   *a.* cardiac muscle
   b. smooth muscle   *c.* skeletal muscle   *d.* voluntary muscle.

Matching

6. synovial fluid
7. ligaments
8. tendons
9. sarcomeres
10. endoskeleton

a. units that make up myofibrils
b. hold bones together
c. hold muscle to bones
d. protects the ends of bones
e. bones found inside a protective covering

Fill In

11. The changing of cartilage to bone is called _____.
12. There are about _____ bones and _____ muscles in the human body.
13. Cardiac muscle is found only in the _____.
14. The muscle used in movement of bones is called _____ muscle.
15. The two proteins found in sliding filaments are _____ and _____.

Portfolio

If your school has a gymnastic team, you might want to take photographs at one of their practice sessions or at a competitive meet. Try to take photographs that show a wide range of body movements. Use these photographs to make a bulletin board display that illustrates the range of movements of bones, joints, and muscles.

Chapter 27
Circulatory System

In ORDER FOR humans to live, they must be able to distribute oxygen and nutrients to every cell in the body and to carry carbon dioxide and waste products away from each of these cells. The **circulatory system** carries out these functions. The circulatory system is made up of a liquid tissue (blood), a network of tubes (the blood vessels), and a pump (the heart).

THE BLOOD

The function of **blood** is to carry nutrients and oxygen to cells and to carry carbon dioxide and other waste products from cells. Blood is a mixture of several things. The liquid portion of

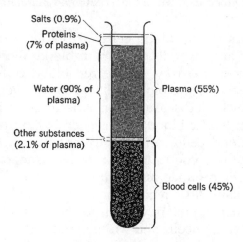

Salts (0.9%)
Proteins (7% of plasma)
Water (90% of plasma)
Other substances (2.1% of plasma)
Plasma (55%)
Blood cells (45%)

Figure 27-1. Composition of human blood.

blood is called **plasma**. Plasma is mostly water and contains many dissolved substances such as proteins, minerals, sugar, and vitamins. In addition to plasma, the blood has solid components—the blood cells. There are three kinds of blood cells: erythrocytes (red blood cells), leukocytes (white blood cells), and thrombocytes (platelets).

Erythrocytes (red blood cells) are made in the red marrow of our bones. Red blood cells are constantly being made, because red blood cells live for only about four months before they die and are removed from the circulatory system by the spleen. Red blood cells contain a red pigment called **hemoglobin.** Hemoglobin molecules contain iron. A hemoglobin molecule has the important job of carrying oxygen to the cells of the body and carbon dioxide away from the cells. Red blood cells are shaped like a doughnut without the hole. This "double concave" shape enables them to carry more oxygen.

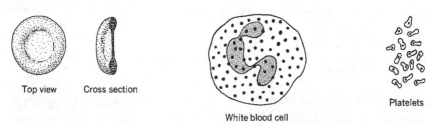

Top view   Cross section

White blood cell

Platelets

Figure 27-2. Red blood cell, white blood cell, and platelets.

Leukocytes (white blood cells) are larger than red blood cells, and there are fewer of them in our blood. White blood cells are made in bone marrow. Unlike red blood cells, leukocytes may live for years. White blood cells perform several vital functions. First, they fight disease. When a disease-causing organism enters the body, white blood cells gather around the organism. Some white blood cells can actually engulf unwanted organisms. These white blood cells are called **phagocytes.** Other white blood cells produce **antibodies,** substances that will destroy the organism. After a battle between white blood cells and invading organisms, pus may begin to form. **Pus** is a combination of dead invading cells and white blood cells.

Thrombocytes (platelets) are fragments of cells that are released from marrow. Platelets play an important role in causing blood to clot. If a blood vessel tears, platelets start a series of chemical reactions that result in the formation of **fibrin.** Fibrin is a network of threadlike material that plugs the torn area. Large red blood cells become trapped in the fibrin and harden, forming a clot or a scab.

BLOOD TYPES

Other important components found in our blood are the **antigens**. Antigens are located on the surface of red blood cells. They help in the production of antibodies. Antibodies help us fight disease. Scientists use antigens to classify our blood in groups known as blood types.

There are four main **blood types:** A, B, AB, and O. If red blood cells have antigen A on their surface, then the person has type A blood. If antigen B is present, the person has type B blood. If both A and B antigens are present, the individual has type AB blood. If neither A nor B antigen is found, the person has type O blood.

Because of the different antigens on the surface of red blood cells, some blood types do not mix with other blood types. Instead the blood clumps, or **agglutinates.** When a transfusion of blood is needed, it is important to make sure the blood types of the donor and the recipient match. Type O blood is called the **universal donor.** Because there are no antigens present, O blood can be given to any other blood group and the blood will

Table 27-1. Blood Types and Compatibilities

Type	Antigen on Red Blood Cells	Antibodies in Plasma	Can Receive	Can Give to
A	A	b	A or O only	A or AB only
B	B	a	B or O only	B or AB only
AB (universal recipient)	AB	None	All four types	AB only
O (universal donor)	None	a and b	O only	All four types

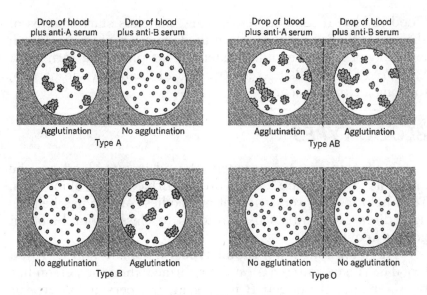

Figure 27-3. Clumping of different blood types.

not agglutinate. Type AB blood is called the **universal recipient.** People with AB blood can receive blood from people with types A, B, AB, and O without agglutination occurring. Receiving incompatible blood can produce very serious consequences. Clumping will occur in the blood vessels, and might interfere with the normal flow of blood.

THE RH FACTOR

Another important antigen found in humans is called the **Rh factor.** Most people in the United States have this antigen and are Rh-positive. If you don't have this antigen present in your blood, you are Rh-negative. The Rh factor is usually indicated by a plus or minus sign shown after your blood type, such as A+, or AB-. The Rh factor must also be matched during blood transfusions so that agglutination does not occur.

Sometimes a problem that involves the Rh antigen occurs during pregnancy. When the baby is Rh+ and the mother is Rh–, the mother is carrying a baby with an antigen that she does not have. If, during childbirth, some of the baby's blood enters the mother's bloodstream, the mother produces antibodies against the Rh+ antigen. Once the mother has these antibodies in her blood, her body could use them to attack Rh+ antigens of future babies. This could cause the woman to miscarry her next baby. This condition is called **erythroblastosis fetalis.** Physicians are now able to prevent this condition from occurring. After the birth of an Rh+ baby, physicians inject the Rh- mother with a substance that destroys any antibodies she has produced against the Rh+ antigen. This procedure insures that the next baby the woman carries is protected.

ARTERIES, VEINS, AND CAPILLARIES

Blood, with all its components, is carried through the body in a system of tubes called **blood vessels.** There are three types of blood vessels found in humans: arteries, veins, and capillaries. Each type is responsible for carrying blood in a specific direction. Arteries carry blood away from the heart. Veins carry blood toward the heart. Capillaries connect the arteries and veins. The blood makes a continuous circuit through the body.

Arteries carry blood pumped from the heart. These blood vessels have a thick muscle layer, so that the tubes can withstand the tremendous amount of pressure applied when the heart contracts. All arteries, except the pulmonary artery, carry blood that is rich in oxygen, or **oxygenated**. As the arteries pass through the body, they branch into smaller tubes called **arterioles.** The walls of an artery are flexible and "pulse" as blood passes through them. You can feel blood as it flows through an artery at your wrist. In some individuals, the walls of the arteries become

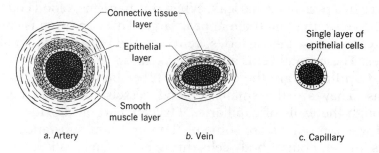

Figure 27-4. Cross sections of an artery, a vein, and a capillary.

inelastic and are unable to flex as the blood is pumped through them. This disease is called **arteriosclerosis.**

Veins usually carry **deoxygenated** blood that needs to pick up oxygen. The pulmonary vein is the only vein that carries blood that is rich in oxygen. Veins have thinner walls than arteries and are not as elastic. Because veins take blood back to the heart, the pressure of the blood is low. In fact, it may be so low

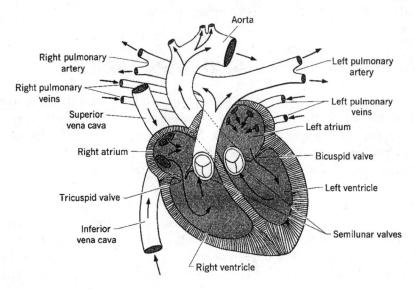

Figure 27-5. The human heart.

that the blood could flow in the wrong direction, away from the heart. To prevent this backward flow, veins have **valves.** These valves can open only in the direction of the heart. Sometimes the valves in a person's veins leak. When this happens, blood collects in the veins, making them appear large. This condition is called **varicose veins.** Like arteries, veins tend to branch into smaller tubes. These small veins are called **venules.**

Capillaries are the connecting tubes between arteries and veins. They are the smallest blood vessels. Diffusion occurs through the walls of capillaries. Oxygen, carbon dioxide, food, and waste products are some of the important substances that pass into and out of body cells through capillary walls.

THE HEART

The **heart** is the pump that is responsible for moving blood through the blood vessels. The heart is made of **cardiac muscle.** Cardiac muscle contracts, pumping blood throughout the body. Signals from the brain stimulate the heart to contract.

The heart is located in the chest between the lungs and behind the sternum, or breastbone. This location offers some protection from outside injury. A membrane sac, the **pericardium,** surrounds and helps protect the heart. The pericardium also produces a fluid that lubricates the outside of the heart.

The human heart has four chambers. The top two chambers are called **atria,** and the bottom two are called **ventricles.** The atria receive blood from the body, and the ventricles pump blood to the body.

There is a specific route that blood takes through the heart. Deoxygenated blood from the body flows through the superior and inferior **vena cava** (large veins) and enters the right atrium. The blood then passes through a **tricuspid valve** and enters the right ventricle. From the right ventricle, the blood is pumped out through the **pulmonary arteries** to the lungs, where oxygen is added to the blood. Then the blood returns to the heart to be pumped to the rest of the body.

The oxygenated blood returns to the heart through the **pulmonary veins.** The pulmonary veins empty the blood into the left atrium. The blood then passes through a bicuspid valve to the left ventricle. When the left ventricle contracts, the oxygen-rich blood is pumped to the body through a large artery called the **aorta.** The path the blood takes through the heart makes sure that deoxygenated blood is always on the right side of the heart and that oxygenated blood is always on the left side of the heart.

When blood moves throughout the body, it is called **circulation.** Doctors often refer to specific areas of circulation. The blood that flows to and away from the lungs is referred to as **pulmonary circulation.** The blood that passes through vessels in the kidneys is called **renal circulation. Coronary circulation** supplies the heart with blood. **Systemic circulation** is the flow of blood to all the organs of the body except the lungs.

The Lymphatic System

The **lymphatic system** is the part of the circulatory system responsible for collecting excess tissue fluid. Tissue fluid is the liquid that is found around all cells of the body. Open-ended **lymph vessels** collect this excess fluid. Once the fluid is inside lymph vessels, it is called **lymph.** Some of the lymph vessels join veins near the heart. Here the lymph becomes part of the blood. **Lymph nodes** are filters found in the lymphatic system. Collections of lymph nodes are located in the neck, groin, and armpits. Any bacteria, foreign particles, and parts of cells are filtered from the blood in the lymph nodes and destroyed. The lymph nodes also produce cells that fight disease. When you are ill, the lymph nodes often become enlarged.

QUESTIONS

Multiple Choice

1. The liquid portion of the blood is called the   a. erythrocyte   b. plasma   c. thrombocyte   d. platelets.

2. The protective sac that surrounds the heart is the   a. pericardium   b. ventricle   c. lymph   d. atrium.

3. The antigen involved in erythroblastosis fetalis is   a. A   b. B   c. AB   d. Rh.

4. The flow of blood to and from the lungs is called the   a. systemic circulation   b. renal circulation   c. lymphatic circulation   d. pulmonary circulation.

5. Which of the following is responsible for carrying oxygen in the blood?   a. platelets   b. thrombocytes   c. erythrocytes   d. leukocytes

Matching

6. atrium	*a.* lower chamber in heart
7. ventricle	*b.* large artery
8. tricuspid valve	*c.* valve on right side of heart
9. bicuspid valve	*d.* top chamber of the heart
10. aorta	*e.* valve on left side of heart

Fill In

11. The smallest type of blood vessel is the _____.

12. The _____ blood type is known as the universal donor.

13. _____ are the blood vessels with valves that prevent the backward flow of blood.

14. The blood vessels that carry blood away from the heart are the _____.

15. The circulation of blood to all the organs of the body except the lungs is called _____ circulation.

Portfolio

Like all muscles in the body, the heart can be affected by exercise. Find out the kinds of exercises that benefit the heart. Make drawings or photographs that show people performing these exercises. Use this information to make a bulletin board display, or a presentation for your class, or to write an article for your school newspaper.

Chapter 28

Respiratory System

THE HUMAN RESPIRATORY system is essential to life because we need a constant flow of oxygen to each cell. This specialized system is also responsible for removing some waste materials from our bodies. It is important to understand the organs that make up the respiratory system, their function, and how we can keep them working efficiently.

RATE OF BREATHING

If you run around a city block, or up a steep hill, or in a long race, you will notice changes in your body. One important change occurs in your rate of breathing. Your breathing rate speeds up with strenuous exercise because your body needs to take in more oxygen and get rid of more carbon dioxide.

Our respiratory system removes oxygen from the air in direct proportion to our body's needs. As we begin to exercise or take part in strenuous activity, the number of breaths we take per minute increases to accommodate our body's need for increased amounts of oxygen. When we are asleep, our cells require less oxygen, and our breathing rate slows. This rate is automatically controlled by the respiratory center in a part of the brain called the **medulla oblongata.** Although we can control our breathing rate for a short amount of time, by holding

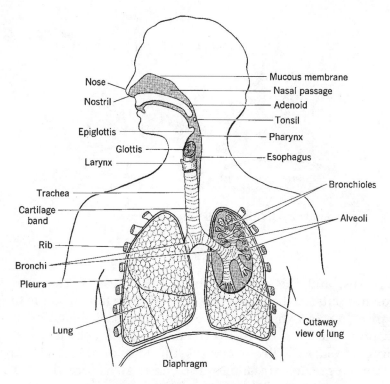

Figure 28-1. Structures of the human respiratory system.

our breath or by deliberately taking deeper or shallower breaths, eventually our brain takes control of our respiratory system. Someone who purposely holds her or his breath will eventually become unconscious. While unconscious, the brain adjusts the body to a normal breathing pattern.

THE PATH OF A BREATH

The **diaphragm,** located just beneath the lungs, is a thick sheet of muscle that divides the abdomen and the chest. The diaphragm is responsible for controlling the mechanism of breathing. **Breathing** is the process of taking air into our lungs and then forcing air out of our lungs. Breathing is divided into two steps. **Inspiration,** or inhalation, is the process of taking air into our lungs. When we inhale, the diaphragm contracts, causing the ribs to move up and out. The size of the chest cavity

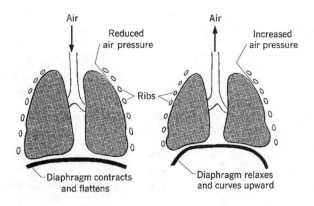

Figure 28-2. Inhalation and exhalation.

increases, and the pressure inside the chest is less than the pressure outside the body. This causes air to flow into the lungs. **Expiration,** or exhalation, is forcing air out of the lungs. In expiration, the ribs move down and inward because the diaphragm relaxes. This decreases the size of the chest cavity, creating a greater pressure in the chest. Air is forced out of the body from this area of greater pressure.

Respiration begins as air enters our **nasal passages,** or occasionally our mouth. When air is taken in through the nose, it passes through the **nostrils,** where hairs filter dust and other particles from the air. The air also passes over a layer of mucus. Particles of dust, smoke, and even bacteria stick to the mucus and are removed from the stream of inhaled air. The nasal cavity is also lined with **cilia,** tiny hairs that sweep unwanted substances into the throat where they are coughed out or swallowed. The small blood capillaries in the nose warm and moisten the air.

The air then passes by the **pharynx,** an opening at the back of the mouth. The pharynx is a common passageway for both the respiratory system and the digestive system.

Since both food and air pass by the pharynx to the **trachea,** or windpipe, it is important to keep food or liquids from entering the trachea and blocking the flow of air. When you swallow, a flap called the **epiglottis** covers the trachea's opening and keeps material from entering. When you breathe, the epiglottis does not cover the opening of the trachea. Air passes freely into the trachea.

The trachea is about 25 centimeters long. It is lined with cilia, which act as another filter for the respiratory system. The trachea is strengthened by rings of cartilage. These rings keep the trachea from collapsing. The **larynx,** or voice box, is located at the upper end of the trachea. The larynx contains the **vocal cords.** The vocal cords are responsible for producing our voice.

The trachea eventually branches into two **bronchi**. Each bronchus leads to a lung. The **lungs** are the major organs of the respiratory system. The bronchi then branch into even smaller and smaller tubes inside the lungs. The smallest tubes are called **bronchioles.** The bronchioles end in tiny air sacs called **alveoli.** Each lung contains several hundred million of these tiny air sacs. A network of capillaries surrounds each alveolus.

At the alveoli, external respiration—the exchange of oxygen and carbon dioxide—occurs. The walls of these tiny air sacs are thin and moist, making it easy for the exchange of gases. **External respiration** is the exchange of gases between the atmosphere and the blood. Oxygen moves into the blood, and carbon dioxide is taken out of the blood. These gases are also exchanged between the blood and the cells of the body. This process is called **internal respiration,** or cellular respiration.

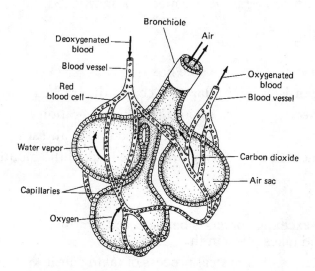

Figure 28-3. Alveolus.

The respiratory system provides the body with an efficient way to transport gases to and from the circulatory system. Working together, the circulatory and respiratory systems are able to provide the oxygen needed to continue the important process of cellular respiration.

QUESTIONS

Multiple Choice

1. The tiny air sacs in the lungs are called   *a.* pleura *b.* vocal cords   *c.* alveoli   *d.* receptors.
2. The tubes that carry oxygen to the lungs are the *a.* alveoli   *b.* bronchioles   *c.* tracheas   *d.* pleura.
3. Which of the following muscles is important in the process of breathing?   *a.* sternomastoid   *b.* biceps   *c.* latissimus dorsi   *d.* diaphragm
4. When a person breathes in, the ribs move   *a.* up and in *b.* down and in   *c.* up and out   *d.* down and out.
5. In internal respiration, gases are   *a.* exchanged between the blood and cells   *b.* exchanged between the lungs and the atmosphere   *c.* warmed   *d.* moistened.

Matching

6. pharynx   *a.* voice box
7. epiglottis   *b.* flap that covers the trachea
8. larynx   *c.* major organ in respiration
9. lung   *d.* small tubes ending in air sacs
10. bronchioles   *e.* opening at the back of the mouth

Fill In

11. The exchange of gases between the atmosphere and the blood takes place in the _____.
12. _____ is the physical process of taking air into the lungs and removing air from the lungs.

13. The tiny hairs that help filter dirt from the nasal area are called _____.

14. Breathing is controlled by a region in the brain called the _____.

15. The diaphragm is a muscle located beneath the _____.

Portfolio

Smoking tobacco products introduces harmful chemicals into the respiratory system. Find out what some of these chemicals are and the ways they affect body tissues. Write a brief report on the effects of smoking on the human body.

Chapter 29
Digestive System

DID YOU EVER wonder how the foods you eat supply energy for your body? Or how the foods you eat "become" you? The human digestive system is responsible for converting the foods we eat and drink into smaller molecules that can be used for energy, growth, and repair of body tissues.

The foods we eat are made of nutrients. **Nutrients** are the basic building blocks living things need to live and grow. The nutrients we eat or drink are proteins, fats, carbohydrates, vitamins, minerals, and water. All foods are made from these six groups of nutrients. However, few foods contain all of the nutrients our body needs. Therefore, it is important to eat a wide variety of different foods for good health.

IMPORTANT NUTRIENTS

Proteins are used to build important structures in cells. Proteins are essential for growth and repair of body tissues. One group of proteins, the **enzymes,** are the major organic catalysts in the body. They take part in many important chemical reactions.

Fats can be used by the body for energy or can be stored as energy reserves. The body also uses fats to build membranes and other important cell parts. Physicians recommend that you limit the amount of fats you eat, because they contain so many calories. Fats are also considered to be contributing factors in certain diseases.

Carbohydrates provide most of the energy needed by the body. Sugars and starches are carbohydrates—compounds made up of carbon, hydrogen, and oxygen. These compounds are easily converted into energy by the body. Proteins and fats can also be used by the body for energy, but the chemical structures of these compounds make it more difficult for the body to convert them to energy.

Vitamins are molecules that are involved, along with enzymes, in many metabolic activities. Vitamins can be absorbed without being digested. With the exception of vitamin D and vitamin K, vitamins needed for the proper functioning of the human body must be supplied by foods or supplements.

Minerals are inorganic elements needed for normal functions of the body. Iron, calcium, and iodine are some of the minerals the body needs. The human body needs only small amounts of most minerals. Minerals can be found in many vegetables.

Water makes up more than half of your body weight. Water is essential for many of the activities that occur in the body, since most chemical reactions take place in water. We also need water to help us regulate our body temperature. The water we naturally lose each day must be replaced.

THE DIGESTIVE SYSTEM

Digestion is a series of physical and chemical processes that break food down into smaller bits our body can use. Any food we eat follows the same path and process until it becomes usable by our bodies. The trip begins as soon as we put a bite of food in our mouth.

Because humans are omnivores and eat a variety of foods, our teeth are shaped to physically cut, shred, chew, and grind our food into smaller chunks. In the mouth, saliva moistens the food and begins chemical digestion.

Saliva is a mixture of mucus, water, and a digestive enzyme known as salivary amylase. Saliva is produced by salivary glands located in the lining of the mouth. The chunks of chewed food moistened with saliva are known as a **bolus,** or ball. This bolus is forced by the tongue down toward the back of the throat.

Table 29-1. Nutrient Summary

Nutrient	Composition	Uses in the Body	Rich Sources	Chemical Test
Carbohydrates (sugars and starch)	Carbon, hydrogen, and oxygen; ratio of hydrogen atoms to oxygen atoms is 2:1	The chief source of energy; excess carbohydrates are stored for future use	Sugars: fruit, honey, candy, ice cream Starch: cereal grains, potatoes, bread	Cover with Benedict's solution; heat strongly; orange-to-red color indicates glucose Add iodine solution; blue-black color indicates starch
Lipids (fats and oils)	Carbon, hydrogen, and oxygen; ratio of hydrogen atoms to oxygen atoms is not 2:1	A source of energy; excess fats and oils are stored for future use	Butter, lard, cream, bacon, meat, olive oil, nuts	Grease spot on unglazed paper indicates fats
Proteins	Nitrogen in addition to carbon, hydrogen, and oxygen; some proteins also contain sulfur, iron, phosphorus, and other elements	Provide materials for assimilation and for growth and repair of all body cells; can supply energy	Milk, cheese, eggs, beef, liver, fish, peas, beans, nuts	Cover with dilute nitric acid; heat gently; yellow color develops; pour off acid; cover with ammonia water; orange color indicates protein

Nutrient	Composition	Uses in the Body	Rich Sources	Chemical Test
Water	Hydrogen and oxygen	Makes up more than 70 percent of protoplasm; it is the solvent in which the chemical reactions of protoplasm occur	Drinking water, milk, and other beverages; most foods, particularly fruits and vegetables	Heat food in test tube; condensed droplets on glass indicate water; verify with blue cobalt chloride paper, which turns pink in presence of water
Mineral salts	Chiefly calcium, phosphorus, iron, iodine, fluorine, sodium, and chlorine	Calcium and phosphorus make up hard parts of bones and teeth	Milk, eggs, cheese	Heat strongly in fireproof container until contents glow; allow to cool; white ash indicates minerals
		Iron makes up part of hemoglobin and cytochrome enzymes	Liver, eggs, beef, green vegetables	
		Iodine constitutes a large part of the secretion of the thyroid	Sea foods, iodized table salt	

(Continued on next page)

Table 29-1. Nutrient Summary (continued)

Nutrient	Composition	Uses in the Body	Rich Sources	Chemical Test
Mineral salts (continued)	Chiefly calcium, phosphorus, iron, iodine, fluorine, sodium, and chlorine	Fluorine makes tooth enamel hard and resistant to decay	In some natural water, but often added	Heat strongly in fireproof container until contents glow; allow to cool; white ash indicates minerals
		Sodium takes part in the life functions of nerve and other cells	Table salt	
		Chlorine is a part of hydrochloric acid, necessary in digestion	Table salt	
Vitamins	Carbon, hydrogen, oxygen, nitrogen, and other elements	See Table 29-2		

Table 29–2. Vitamin Summary

FAT-SOLUBLE VITAMINS		
Vitamin	*Sources*	*Result of Deficiency*
Vitamin A (carotene is converted to vitamin A in the body)	Carrots and other yellow vegetables; whole milk, butter, eggs; leafy green vegetables, peas; fish-liver oils	Xerophthalmia, or "dry-eye" (an eye infection); night blindness (inability to see in dim light); increased susceptibility to infections of the nose, throat, and skin
Vitamin D (calciferol)	Fish-liver oils; milk, eggs (Not commonly found in foods. Some foods are irradiated to increase their vitamin D content. If directly exposed to sunlight, the body is able to manufacture vitamin D. For this reason it is called "the sunshine vitamin.")	Rickets (bow-legs, knock-knees, swollen joints, especially wrists and ankles); badly formed teeth
Vitamin E (tocopherol)	Wheat germ; leafy green vegetables; meat; whole grain cereals	Sterility (inability to reproduce) in rats, and possibly in humans
Vitamin K	Leafy green vegetables	Hemorrhage (excessive bleeding, even from minor wounds; the body is unable to make prothrombin, a protein necessary for the normal clotting of blood)

(Continued on next page)

Table 29–2. Vitamin Summary (*continued*)

	Vitamin	Sources	Result of Deficiency
WATER-SOLUBLE VITAMINS			
B-complex	Vitamin B_1 (thiamin)	Whole grain cereals and enriched bread; beans and peas; yeast; milk; lean beef, liver, pork, poultry	Loss of appetite; limited growth; improper oxidation of foods, especially of carbohydrates; beriberi (exhaustion, paralysis, heart disease)
	Vitamin B_2 or G (riboflavin)	Yeast; lean beef, liver; milk, eggs; green vegetables; whole wheat; prunes	Inflamed lips; general weakness; eyes excessively sensitive to light (deficiency of riboflavin is often associated with pellagra)
	Niacin	Lean beef, liver; milk, eggs; leafy green vegetables, tomatoes; yeast	Pellagra (skin irritation, tongue inflammation, digestive and nervous disturbances)
	Vitamin B_{12} (cobalamin)	Liver, kidney; fish	Pernicious anemia; retarded growth; disorders of the nervous system
	Folic acid	Leafy green vegetables; yeast; meat	Some types of anemia
	Vitamin C (ascorbic acid)	Citrus fruits (oranges, grapefruits, lemons, limes); leafy green vegetables, tomatoes	Scurvy (soft and bleeding gums, loose teeth, swollen and painful joints, bleeding under the skin)

The **pharynx** is a common passageway where food passes to the esophagus and air passes to the lungs. The bolus of food is prevented from entering the windpipe and lungs by a flap of tissue called the epiglottis. Food is routed to the **esophagus,** a long muscular tube that leads from the throat to the stomach. Once in the esophagus, the food is moved down the tube by the rhythmic

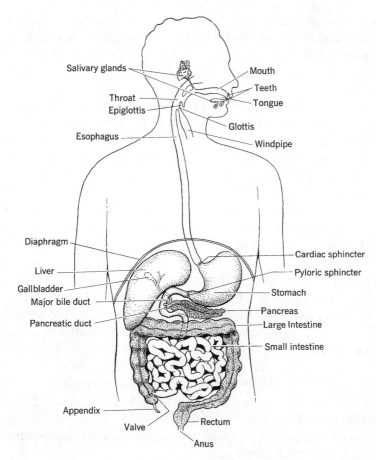

Figure 29-1. The human digestive system.

contractions of the muscles that make up the walls of the esophagus. These rhythmic contractions are called **peristalsis.**

At the end of the esophagus, the bolus of food passes through an opening called the **cardiac sphincter** into a muscular organ called the **stomach.** The stomach is a J-shaped sac that should comfortably hold two handfuls of food. The stomach acts as a blender, mixing food while gastric fluids work to dissolve and digest the bolus. The **gastric fluids** are produced by gastric glands found in the lining of the stomach. These fluids are a mixture of enzymes, mucus, and hydrochloric acid.

The food is physically mixed or churned for two to three hours. During this time it becomes a juicy, soupy mixture called

chyme. The chyme is then moved—in small amounts—out of the stomach and into the small intestine through an opening called the **pyloric sphincter.** In the small intestine, the chyme is acted on by additional chemical agents secreted by the liver and the pancreas.

The **liver** is a large organ that lies in the upper right side of the body cavity. It is the largest gland in the body. The liver secretes a substance called **bile** that is vital to the digestion of fats. Bile is not an enzyme, but it breaks fats up into small droplets that can be acted upon by enzymes present in the small intestine. Bile is stored in a small organ called the **gallbladder,** which lies beside the liver. When chyme enters the small intestine, the gallbladder releases bile into the intestine.

The **pancreas** is a slender gland located beneath the stomach. The pancreas secretes a liquid called pancreatic juice, which is emptied into the small intestine. Pancreatic juice contains several enzymes that break down starches, fats, and proteins.

The **small intestine** is divided into three sections—the duodenum, jejunum, and ileum. The word small refers to the narrow diameter of this tube, not its length. In length, your small intestine is more than three times your height. It is coiled and curved to fit in your abdomen. The digestive juices from the liver and pancreas enter the small intestine in the section known as the **duodenum.** It is here that these juices act on the chyme. The food then passes through the **jejunum** and the **ileum** sections of the small intestine.

The lining of the entire small intestine has many folds that are covered with small projections called **villi.** The villi have their own projections called **microvilli.** The folds, villi, and microvilli increase the surface area of the small intestine. The increased surface area allows more area for absorption. **Absorption** takes the products of digestion and transfers them to the circulatory system where they can be delivered in the blood to all the cells of the body.

The **large intestine,** or **colon,** is the last major organ of digestion. The large intestine's job is to collect the waste products of digestion. These waste products are called **feces.** The colon is divided into three sections—the **ascending, transverse,** and **descending** colon. The final part of the large intestine is the

rectum. The rectum opens to the outside of the body through the **anus.** This opening allows us to rid the body of waste products produced during the process of digestion.

QUESTIONS

Multiple Choice

1. The flap that covers the windpipe to prevent food from entering it is the   *a.* duodenum   *b.* peristalsis   *c.* bolus   *d.* epiglottis.

2. Enzymes are very important catalysts in the body. They are classified as   *a.* lipids   *b.* proteins   *c.* carbohydrates   *d.* nucleic acids.

3. The liver produces a substance that helps the body in the process of digestion. This substance is called   *a.* bolus   *b.* pancreatic juice   *c.* gastric juice   *d.* bile.

4. The folds in the small intestine are covered by small projections called   *a.* chyme   *b.* villi   *c.* sphincters   *d.* pharynx.

5. The muscle contractions that occur in the esophagus are called   *a.* villi   *b.* peristalsis   *c.* bolus   *d.* chyme.

Matching

6. colon
7. liver
8. stomach
9. pharynx
10. pancreas

a. largest gland in the body
b. common passageway for both digestive and respiratory systems
c. large intestine
d. located beneath the stomach
e. J-shaped muscular organ

Fill In

11. Bile secreted from the liver is stored in the _____.
12. The tube that carries a bolus from the mouth to the stomach is the _____.

13. The duodenum, jejunum, and ileum are sections of the
    ———.

14. ——— are a source of quick energy for the body.

15. The mixture of enzymes, acid, and mucus secreted in the
    stomach is called ——— fluids.

Portfolio

Keep a record of all the foods you eat for a week. Find out
what nutrients were in the foods you ate. Did you include a
variety of foods? Did you eat a balanced diet with the correct
amounts of nutrients?

Chapter 30

Excretory System

Every moment of your life—when you are asleep and when you are awake—chemical reactions are going on in your body's cells. Human body functions create many wastes that our body systems dispose of in a variety of ways. For example, our lungs remove carbon dioxide from our blood, which we then exhale into the air. The skin serves as an excretory organ when the sweat glands excrete salts, water, and urea. Our digestive system removes all the nutrients we need from our food, and then we eliminate indigestible matter and waste products as feces. **Excretion** is the process by which wastes are removed from the body. The skin, lungs, and kidneys—and other organs associated with them—make up the human excretory system.

THE KIDNEYS

The **kidneys** are our primary tools for the removal of chemical wastes from our system. It is important that these wastes are removed from the blood, because they are useless and potentially poisonous if they remain in the blood. Our kidneys are shaped much like the familiar kidney beans. There are two kidneys in the body, each about the size of your clenched fist. The kidneys are located in the lower back, one on each side of the spine. Each kidney has two distinct regions that perform different functions in the waste removal process.

219

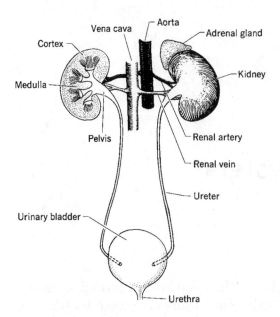

Figure 30-1. The human urinary system.

The **renal cortex** is the outer region of the kidney. **Nephrons,** the most important structures in the kidney, are located in the cortex. The nephron is the actual site of blood filtration. Each kidney has at least one million nephrons. Each nephron is composed of a Bowman's capsule, a glomerulus, and renal tubules.

Blood enters the kidney through a branch of the aorta called the **renal artery.** As the renal artery enters the kidney, it gets smaller and smaller until it eventually forms a net of capillaries called the **glomerulus.** Each glomerulus is surrounded by a cuplike structure called **Bowman's capsule.** The pressure on the blood in the capillaries causes the liquid portion of the blood, the **filtrate,** to leave the glomerulus and enter Bowman's capsule. The filtrate contains a variety of substances, such as water, proteins, carbohydrates, and waste products. The filtered blood later leaves the kidney by way of the **renal vein.**

The filtrate moves inside a series of small tubes in the nephron called **tubules.** As the filtrate moves through the nephron, useful substances are reabsorbed into the blood, and waste products continue to travel through the nephron. The tubules located closest to Bowman's capsules are called **proximal tubules,** and the ones located the farthest from the capsules are called **distal tubules.** The proximal and distal tubules are connected in an

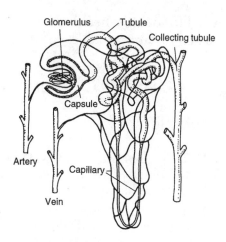

Figure 30-2. A nephron.

area known as the **loop of Henle.** This loop actually dips down into the **renal medulla,** the inner region of the kidney.

The distal tubules drain into the **collecting duct.** Liquid that enters the collecting duct is called **urine,** and it contains urea, salts, and excess water. The collecting duct passes through the medulla and empties into the **renal pelvis,** a large cavity in which urine collects. Sometimes minerals in the renal pelvis crystallize and form **kidney stones,** which can block the flow of urine.

The urine collected in the kidney flows to the bladder through a tube called the **ureter.** Humans have two ureters, one leading from each kidney. The **urinary bladder** is a hollow muscular sac that stores urine until it can be expelled from the body. When the bladder empties, the urine passes through a single tube, the **urethra,** to the outside.

Sometimes disease or injury causes the kidneys to have trouble filtering blood. In severe cases, the kidneys may stop functioning completely. If this occurs, waste products accumulate in the blood. This type of poisoning can result in death.

Some individuals with kidney problems may require a transplant. During a transplant operation, the damaged kidney is replaced by one that functions. Other individuals may need to filter their blood by means of **dialysis,** in which a patient's blood is passed through a machine that filters wastes from the blood.

Even though the excretory system is made up of only a few organs, it is a vital part of a healthy body. In fact, one procedure used to identify health problems is urinalysis, in which urine is tested for abnormal levels of certain chemicals.

QUESTIONS

Multiple

1. The tubes that carry waste products from the kidneys to the bladder are the   *a.* urethras   *b.* ureters   *c.* nephrons   *d.* medullas.

2. The collecting duct empties into a cavity called the renal   *a.* pelvis   *b.* medulla   *c.* cortex   *d.* nephron.

3. The actual site of filtration in the kidney is the   *a.* pelvis   *b.* ureter   *c.* urethra   *d.* nephron.

4. Which of the following stores urine until it is ready to be removed from the body?   *a.* bladder   *b.* pelvis   *c.* urethra   *d.* ureter

5. Filtering of the blood by a machine is called   *a.* reabsorption   *b.* urination   *c.* dialysis   *d.* peristalsis.

Matching

6. loop of Henle	*a.* tubes closest to the Bowman's capsule
7. proximal tubules	
8. distal tubules	*b.* connects two different types of tubules
9. collecting duct	
10. filtrate	*c.* tubes not located near the glomerulus
	d. liquid filtered from the blood
	e. empties into the pelvis

Fill In

11. Kidney _____ can form in the pelvis of the kidney and block the flow of urine.

12. The bed of capillaries located at the end of the renal artery is called the _____.

13. The kidney is divided into two regions, the _____ and the _____.

14. The Bowman's capsule, glomerulus, and tubules are located in the _____ of a kidney.

15. Filtered blood is taken out of the kidney by way of a renal _____.

Portfolio

Does your state include a provision for donating organs on the driving licenses they issue? Write a letter to the Commissioner of Motor Vehicles in your state to find out.

Chapter *31*
Endocrine System

WATCH A DIVER leap from a high diving board and complete a series of complex movements before hitting the water. In the few seconds the dive lasts, the diver must be in precise control of his or her body. Of course, the human body is always "under control," even when completing more ordinary tasks than competitive dives. The human nervous system regulates many body activities. However, another important system helps regulate life functions—the **endocrine system.**

The study of the endocrine system is called **endocrinology.** The changes effected by the endocrine system, however, are more gradual than the changes made by the nervous system.

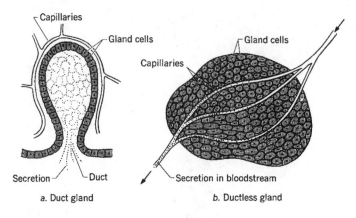

Figure 31-1. The two types of glands.

The endocrine system consists of **endocrine glands** that secrete chemicals directly into the bloodstream. Other glands in the body, **exocrine glands,** release the chemicals they produce through a duct. A tear gland is an example of an exocrine gland.

THE ENDOCRINE SYSTEM

The human endocrine system is made up of several glands located in different parts of the body. The thyroid, parathyroids, adrenals, pituitary, thymus, pancreas, gonads, and pineal are all endocrine glands. These glands send out chemical messengers to all parts of the body.

The chemical messages our endocrine system sends out are called **hormones.** Hormones are carried in the blood so they travel much more slowly than nerve impulses, but their effects on the body usually last much longer. Even though the blood carries the hormones throughout the body, each hormone can influence the action only of **target cells.** However, target cells contain **receptors** for specific hormones. Target cells are often located at a distance in the body from the endocrine glands that produce the hormone that affects them.

The endocrine glands regulate themselves with a type of multistep feedback mechanism. There are two types of feedback mechanisms that work on the human endocrine system. If the end product stops the first step, it is called **negative feedback.** In negative feedback, the end product stops the production of a

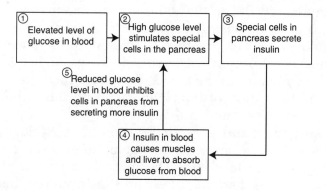

Figure 31-2. Hormonal feedback mechanism.

particular hormone. **Positive feedback** occurs when the end product is needed in order for the first step in the series to begin. Negative feedback is used by most endocrine glands.

ENDOCRINE GLANDS

The endocrine glands secrete hormones directly into the bloodstream. These glands are located in different parts of the body. Some specialized cells also secrete hormones and are considered endocrine "glands."

The **thyroid** is a two-lobed gland located in the neck. The thyroid gland produces the hormone thyroxine. **Thyroxin** helps regulate the body's growth and metabolism (the rate at which the body oxidizes its food). Sometimes the thyroid gland produces too much thyroxin, and an individual develops a condition called **hyperthyroidism.** People with this condition usually have high blood pressure and an overactive metabolism. If the thyroid gland produces too little thyroxin, the individual is said to have **hypothyroidism.** These people tire easily and are often overweight. If a baby has an underactive thyroid gland, he or she may develop a condition called **cretinism.** Cretinism stunts the child's growth and can cause mental retardation. If the condition is diagnosed early, it can be corrected.

Another problem occurs in the thyroid gland when an individual's diet is deficient in iodine. Because the thyroid needs iodine to produce thyroxin, abnormal growth of thyroid tissue can occur, producing an enlarged thyroid called a **goiter.** Goiters are rare today, largely due to the addition of iodine to table salt.

There are four **parathyroid glands** located in the back of the thyroid gland. The parathyroid glands secrete **parathyroid hormone** (parathormone). Parathyroid hormone helps control the levels of calcium in the blood. This hormone is very important in the formation of bones and in controlling the proper functioning of nerves and muscles.

An **adrenal gland** is located on top of each kidney. Each adrenal gland has an inner layer and an outer layer. The inner layer is called the **medulla.** The medulla produces the hormones adrenaline and noradrenaline. **Adrenaline** helps the body respond to stressful situations. It increases the level of

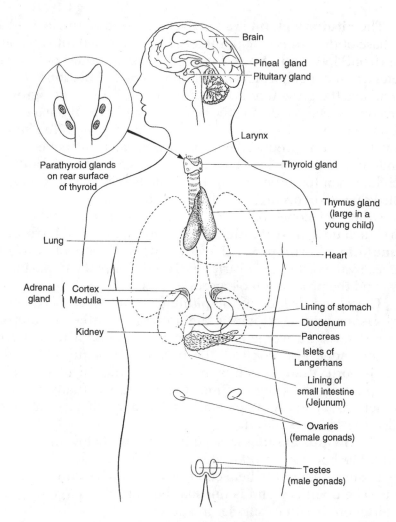

Figure 31-3. The human endocrine system.

blood sugar, increases the heart rate and blood pressure, and enlarges the blood vessels. Adrenaline also increases the flow of blood to the muscles that are attached to the skeleton. **Noradrenaline** produces the same effects on the body as adrenaline.

The outer layer of the adrenal gland is called the **cortex.** The cortex produces corticosteroids and aldosterone. **Corticosteroids,** or corticoids, influence the metabolism of carbohydrates, proteins, and fats. **Aldosterone** helps control the balance of water and salts in the body.

The **pituitary gland** is a relatively small structure located at the base of the brain. The pituitary gland is divided into an anterior (front) lobe and a posterior (rear) lobe. The anterior lobe secretes a number of hormones, including **growth hormone**, which stimulates the growth of bones in the body. If too much growth hormone is secreted during the growth years (up to about age 16), the individual will grow much taller than normal. Too much growth hormone produced in an adult will cause the bones in the hands and face to thicken. This condition is called **acromegaly.** In a child, when too little growth hormone is released, a condition called **dwarfism** occurs. The child remains short.

Other hormones released by the anterior lobe of the pituitary include **LH (luteinizing hormone)** and **FSH (follicle stimulating hormone).** LH causes the secretion of sex hormones by the testes and ovaries. FSH influences the maturing of eggs and the production of sperm.

The posterior lobe of the pituitary gland secretes oxytocin and vasopressin. **Oxytocin** is the hormone that begins uterine contractions in a woman when her child is about to be born. This hormone also controls the production of milk after the child is born. **Vasopressin** is the hormone that regulates water absorption by the kidneys. Since the pituitary actually controls the actions of many other endocrine glands, it is sometimes called the **"master gland."**

The **thymus gland** is located in the chest, below the thyroid gland. The hormone produced by this gland is called **thymosin.** Thymosin is thought to help children develop their immune systems. The thymus gland is unusual because it is large in small children but becomes smaller in adults.

The **pancreas** is found in the abdomen near the stomach. It produces enzymes that are used in digestion. These enzymes are carried in a duct to the small intestine. The pancreas also contains areas of cells that are known as **islets of Langerhans.** The islets produce hormones that enter the bloodstream directly: insulin and glucagon. **Insulin** decreases the level of sugar in the blood by increasing the amount of sugar absorbed by body cells. **Glucagon** increases the level of blood sugar and lowers the amount of stored sugar.

The correct balance of insulin and glucagon insures that the level of sugar in the blood remains within an appropriate range.

If a person lacks the correct amount of insulin, she or he develops **diabetes mellitus.** In this condition, there is a high level of sugar in the blood, and the cells of the body are being denied the sugar they need to produce energy. **Hypoglycemia** occurs when a person has too low a level of sugar in the blood. Hypoglycemia is usually caused when too much insulin is produced.

The **gonads** are the glands responsible for producing gametes. The **testes** in males produce sperm, and the **ovaries** in females produce eggs. In addition to the production of the gametes, testes and ovaries produce hormones. The testes are located in the scrotal sac and produce the hormone testosterone. **Testosterone** produces male secondary sex characteristics, including the growth of facial hair, broadening of the chest, and the deepening of the voice. The ovaries are located in the female's internal pelvic region. The ovaries secrete estrogens and progesterone. **Estrogens** regulate the development of the female secondary sex characteristics. **Progesterone** helps control the female's monthly reproductive cycle.

The **pineal gland** is located deep inside the brain. It produces **melatonin,** which may help regulate the onset of puberty and sexual maturity.

The endocrine system plays many important roles in regulating the body's growth and development. Endocrinology is an important area of study, as scientists continue to discover new hormones and further develop their understanding of our body functions and development.

QUESTIONS

Multiple Choice

1. Insulin is secreted by the   *a.* pineal gland   *b.* thyroid gland   *c.* islets of Langerhans   *d.* thymus gland.

2. Which of the following glands are important in a time of emergency?   *a.* parathyroid   *b.* adrenal   *c.* pituitary   *d.* thymus

3. Which of these glands could be the cause of high blood pressure and an overactive metabolism?   *a.* thyroid   *b.* parathyroid   *c.* pineal   *d.* thymus

4. The main result of the release of insulin is   *a.* an increase in the level of blood sugar   *b.* the production of antibodies *c.* a decrease in the level of blood sugar   *d.* a lowering of calcium in the blood.
5. Which of the following glands is called the master gland? *a.* pineal   *b.* thyroid   *c.* thymus   *d.* pituitary

Matching

6. thyroid   *a.* adrenaline
7. pituitary   *b.* insulin
8. adrenal   *c.* thyroxin
9. pancreas   *d.* growth hormone
10. thymus   *e.* thymosin

Fill In

11. Too much thyroxin results in a condition called _____.
12. The study of the endocrine system is called _____.
13. _____ are the chemical messengers of the endocrine system.
14. Two hormones that help maintain the correct level of sugar in the blood are _____ and _____.
15. Both the thyroid and _____ glands are located in the neck region.

Free Response

16. Why do you think the pituitary is often called the "master gland"?
17. Name the hormones that help regulate the sugar level in the body. Tell where they are produced and what they do.
18. What are hormones, and why are they important in maintaining a healthy body?
19. How could a lack of receptors result in the same symptoms as a missing hormone?
20. What are some similarities and differences between cretinism and dwarfism?

Chapter 32

Nervous System

RIDE A BICYCLE. Fly a space shuttle. Recognize a bird's song. Memorize the words to a poem. Think about solving a mathematics problem. All of these complex tasks can be accomplished with the help of your nervous system.

The human **nervous system** is a communication system that allows us to monitor our environment and respond to it. It controls and coordinates all our body functions, including our thoughts. The nervous system also sends and receives messages throughout the body.

Our nervous system is divided into the central nervous system and the peripheral nervous system. The **central nervous system** is made up of the brain and spinal cord. Among its many functions, the central nervous system coordinates the activities of the body by reacting to stimuli. The stimuli may originate from inside or outside the body. The central nervous system is also the center of thought. It controls our balance and the perception of our position relative to the world around us. The **peripheral nervous system** conducts messages to and from the central nervous system and the rest of the body.

THE NEURON

The nervous system is made up of billions of specialized cells called **neurons.** Neurons are the largest cells in the body.

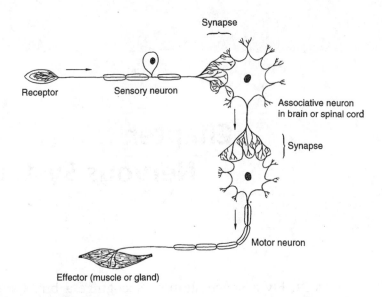

Figure 32-1. Types of neurons.

Unlike other body cells, neurons are not able to reproduce themselves. You are born with all the neurons you will ever have—and in the case of your brain, many more neurons than you will ever need! Neurons carry messages throughout the body, coordinating all body activities. The messages sent through the nervous system are called nerve **impulses.**

There are three types of neurons. **Sensory neurons** receive stimuli and transmit messages to the brain and spinal cord. Clusters of sensory neurons are found in the tips of your fingers,

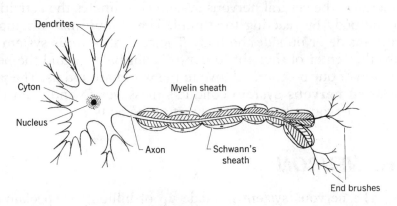

Figure 32-2. Structure of a neuron.

your eyes, nose, and other sensory organs. **Motor neurons** carry impulses from the central nervous system to muscles and glands. **Interneurons,** or **associative neurons,** link sensory and motor neurons.

All neurons have the same basic shape and structure. They have a **cell body,** or **cyton,** which contains a nucleus and provides a place for receiving and sending impulses.

A group of neurons is called a **nerve.** Neurons also have **fibers** that extend from the cell body. **Dendrites** are short, branching fibers that always carry impulses toward the cell body. **Axons** are long fibers that are responsible for carrying impulses away from the cell body. Neurons have only one axon, but it has branches at the end. Some axons have a covering called a **myelin sheath**. The myelin sheath is made of fatty materials that act as insulation and protection for the axon. The myelin sheath is made of cells called **Schwann cells.** The Schwann cells have gaps between them, which are called **nodes of Ranvier.**

THE NERVE IMPULSE

A nerve impulse is a message that travels along a neuron. When no impulse is moving in a neuron, that neuron is said to be in a **resting potential.** During resting potential, there are more sodium ions outside the nerve cell than inside the cell, and more potassium ions inside the cell than outside the cell. Thus, the cell has a positive charge on the outside of the cell membrane and a negative charge on the inside. In this condition, the cell membrane is **polarized.** When the neuron is stimulated, the membrane becomes **depolarized,** and sodium ions rush inside the cell. This results in an **action potential.** The change from a negative to a positive charge inside the cell generates the nerve impulse. As soon as the impulse passes a section of membrane, it returns to a resting potential. A process involving active transport, called the **sodium-potassium pump,** moves sodium ions to the outside of the cell and potassium ions to the inside of the cell. This keeps the sodium and potassium from reaching equilibrium. If the sodium and potassium ever did reach equilibrium, an action potential could not be created and an impulse could not be moved down a neuron.

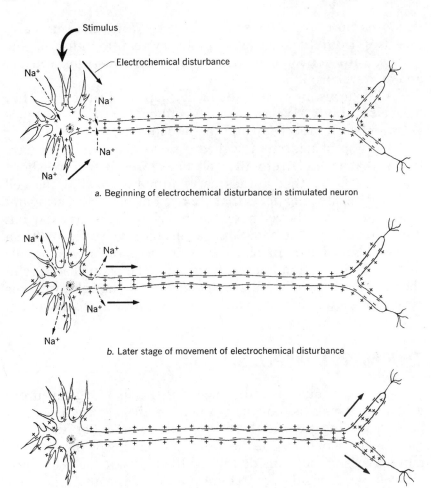

a. Beginning of electrochemical disturbance in stimulated neuron

b. Later stage of movement of electrochemical disturbance

c. Final stage of movement of electrochemical disturbance

Figure 32-3. Movement of a nerve impulse.

SENDING SIGNALS

One of the amazing things about neurons is that they do not actually touch each other, yet they transmit impulses. There is a gap between the axon of one neuron and the dendrite of the next neuron. This gap is called a **synapse.** For many years, scientists wondered how a message could be carried from neuron to neuron across the synapse.

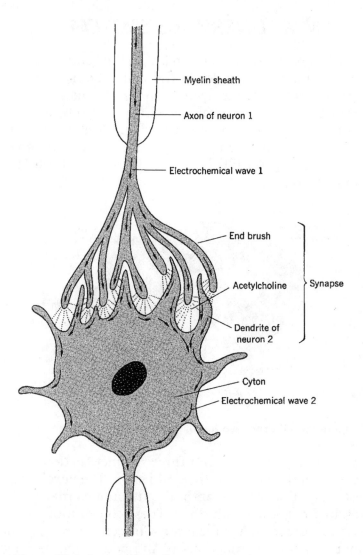

Figure 32-4. A nerve impulse crosses a synapse.

Because neurons do not touch, the impulse must jump across the synapse, or gap. When the impulse reaches the end of an axon, a chemical is released that bridges the synapse. This chemical is called a **neurotransmitter.** It can be released only from **synaptic vesicles** located at the end of the axons. The neurotransmitter diffuses through the synapse and triggers a new impulse on the dendrites of the next neuron.

THE CENTRAL NERVOUS SYSTEM

The brain and spinal cord make up the central nervous system. The brain is surrounded by the skull. The skull's hard, bony case protects the delicate nerve tissue that makes up the brain. The brain, the main control center for the body, contains billions of neurons. Membranes called **meninges** surround the brain, providing protection and nourishment. The space between the meninges is filled with **cerebrospinal fluid.** This fluid cushions the brain.

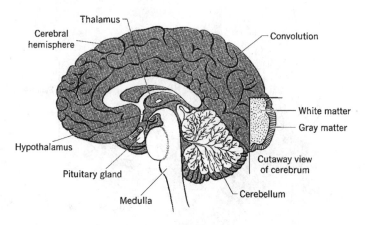

Figure 32-5. Cross section of the brain.

The brain is divided into three distinct sections: the cerebrum, the cerebellum, and the brain stem. The **cerebrum** is the largest section of the human brain. It contains many folds and depressions called **convolutions.** These convolutions greatly increase the surface area of the brain. The cerebrum is responsible for many important functions of the body such as language, thinking, emotions, vision, and hearing. A groove divides the cerebrum into a right and a left **hemisphere.** The left hemisphere controls the right side of the body, and the right hemisphere controls the left side of the body. The hemispheres are further divided into four sections called **lobes.** The four lobes are the **frontal, parietal, temporal,** and **occipital.** Each lobe controls certain functions of the body.

The **cerebellum** is involved in coordinating body movements. Your body's sense of balance is also controlled by the

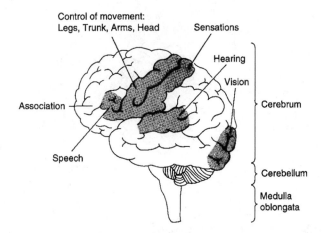

Control of movement:
Legs, Trunk, Arms, Head Sensations

Hearing

Vision

Association

Cerebrum

Speech

Cerebellum

Medulla
oblongata

Figure 32-6. Areas of function in the brain.

cerebellum. The cerebellum is located below the occipital lobe of the cerebrum. In humans, the cerebellum is much smaller than the cerebrum.

The **brain stem** controls important involuntary functions such as heartbeat, blood pressure, and breathing. The brain stem is divided into two regions, the pons and the medulla oblongata. The nerves in the **pons** link the cerebrum and the cerebellum. The **medulla oblongata** controls such automatic responses as heart rate, breathing, blood flow, coughing, and swallowing.

The **spinal cord** connects the brain to the peripheral nervous system. Messages to and from the brain move through the spinal column. The spinal column also controls involuntary movements called **reflexes.** Reflexes allow your body to react quickly to a potentially dangerous situation. The spinal cord is made up of delicate nerve tissue and, like the brain, it is protected by bone. The **vertebrae** make up the backbone, a flexible structure that protects the spinal cord from damage but also permits body movements.

THE PERIPHERAL NERVOUS SYSTEM

The job of this system is to carry messages to and from the rest of the body to the central nervous system. Sensory neurons

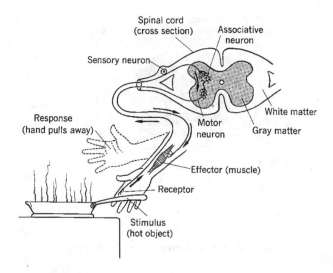

Figure 32-7. A reflex arc.

carry impulses to the central nervous system from receptors located throughout the body. Motor neurons carry impulses from the central nervous system to effectors, which are usually muscles or glands.

The peripheral nervous system can be divided into two separate parts: the somatic system and the autonomic system. The **somatic system** controls movement by controlling the contractions and relaxations of skeletal muscles. The **autonomic system** operates the body's automatic functions such as heartbeat and digestion by controlling involuntary muscles. The peripheral nervous system contains 12 pairs of cranial nerves and 31 pairs of spinal nerves. The **cranial nerves** connect sensory organs and internal organs to the brain. The **spinal nerves** link appendages and the trunk of the body to the brain.

THE SENSES

The job of the **sense organs** is to receive important information from the environment. The eyes, nose, ears, skin, and tongue are all sense organs. Each sense organ contains specialized **receptors** that pick up environmental stimuli. **Photoreceptors** found in the eye respond to light. **Mechanoreceptors** in the skin respond to pressure. **Chemoreceptors** found on the tongue and in the nasal area are able to detect chemicals. **Thermoreceptors** in the skin sense hot and cold. All of these receptors send impulses to the brain to be interpreted.

The **eye** is the organ of vision. Humans are able to focus both eyes on one object. This ability is called **binocular vision,** and it increases depth perception. The eye has many important structures that aid in the process of vision. The outer layer of the eye is called the **sclera.** The sclera is transparent in the front of the eye. This transparent area is called the **cornea.** The cornea protects the eye and permits light to pass into the eye.

The middle layer of the eye is called the **choroid.** This layer includes the **iris,** the colored portion of the eye. The dark opening in the center of the iris is the **pupil.** The pupil can open and close, controlling the amount of light that enters the eye. After light enters the pupil it passes through the lens. The **lens** focuses the light onto the retina. A clear fluid, called the **aqueous humor,** fills the space between the cornea and the lens.

The **retina** is the inner layer of the eye. Receptor cells called **rods** and **cones** are located in the retina. Rods receive impulses from dim light and detect shades of gray. Cones receive impulses from bright light and detect colors. When light stimulates the rods and cones, the light is converted into an impulse made up of an electrical signal. This impulse travels down the **optic nerve** to the brain. At the spot where the optic nerve enters the retina there are no light receptors. This area is known as the **blind spot.** Another clear fluid is located between the lens and

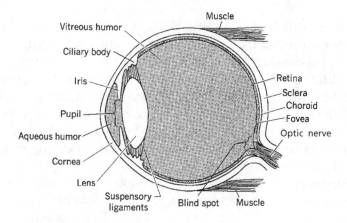

Figure 32-8. Cross section of the eye.

the retina. This liquid is called **vitreous humor** and gives shape
to the eye.

The **ear** is the organ that receives sounds from the environment. Certain structures within the ear also help the body maintain its balance. There is an outer, a middle, and an inner ear.
The outer ear is called the **pinna.** The pinna gathers sound and
conducts it into the **auditory canal.** The auditory canal is a tube
that leads to the **tympanum,** or eardrum. This is a thin membrane that separates the outer and middle ear. The tympanum
vibrates as sound waves come in contact with it.

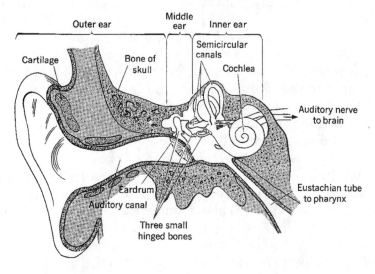

Figure 32-9. Structures of the ear.

The middle ear contains three small bones: the **malleus**
(hammer), the **incus** (anvil), and the **stapes** (stirrup). These
three bones conduct the vibrations from the tympanum through
the middle ear. The stapes touches a membrane that separates
the middle and inner ear. This membrane is called the **oval win-
dow.** There is also a tube in the middle ear that equalizes pressure between the middle ear and the outside of the body. This
Eustachian tube keeps the tympanum from bulging or caving
in due to changes in air pressure.

The receptors for hearing and balance are actually found in
the inner ear. The organ for hearing is the **cochlea.** This fluid-
filled organ is lined with receptor cells. The movement of the
fluid and the stimulation of the receptor cells produce a nerve

impulse. This impulse is carried, by way of the **auditory nerve,** to the brain.

Balance in the human body is achieved by the **semicircular canals** in the inner ear. Stimulation of the cells in the semicircular canals provides information to the brain about the body's position and motion.

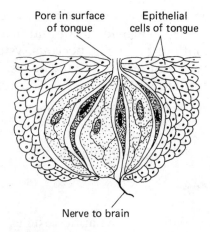

Figure 32-10. A taste bud.

The major organ for taste is the **tongue.** It contains thousands of taste receptors organized into units called **taste buds.** There are about 10,000 taste buds on the tongue. These taste buds are able to distinguish among sweet, sour, salty, and bitter substances.

The organ of smell is the **nose.** The nose contains **olfactory receptors** that pick up chemical molecules in the air. Nerve

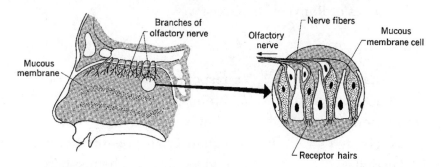

Figure 32-11. Receptors in the nose.

impulses produced at these receptors are carried through the **olfactory nerves** to the brain.

The **skin** contains many receptors capable of detecting a number of different changes in the environment. Touch, pressure, temperature, and pain receptors are found in the skin. Each type of receptor has a different structure and is located in a different area of the skin.

Although the nervous system is very complex, all of its components work together to provide a communication and control network throughout our body. The nervous system is able to initiate, relay, interpret, and respond to messages. Without this important system regulating other systems, the human body could not function.

QUESTIONS

Multiple Choice

1. Nerve cells are called   *a.* synaptic vesicles   *b.* meninges   *c.* Schwann cells   *d.* neurons.

2. The brain and the spinal cord make up the   *a.* peripheral nervous system   *b.* central nervous system   *c.* autonomic nervous system   *d.* somatic nervous system.

3. The area of the eye that contains receptor cells is the   *a.* retina   *b.* pupil   *c.* iris   *d.* cornea.

4. A nerve cell that is resting is said to be   *a.* depolarized   *b.* polarized   *c.* neurotransmitting   *d.* sensory.

5. The long fibers that carry a nerve impulse away from a nerve cell are called   *a.* dendrites   *b.* Schwann cells   *c.* nodes of Ranvier   *d.* axons.

Matching

6. Eustachian tube   *a.* responsible for the sense of smell
7. olfactory receptors   *b.* brain stem
8. humors   *c.* liquids in the eye
9. cerebrum   *d.* equalizes air pressure in the ear
10. medulla   *e.* divided into two hemispheres

Fill In

11. The _____ nerve takes the impulse from the eye to the brain.
12. Being able to focus both eyes on one object is called _____ vision.
13. The brain is protected by several membranes called _____.
14. The gap between two nerve cells is called a _____.
15. The _____ are short branching fibers that carry impulses to a cell body.

Free Response

16. Describe how a nerve impulse crosses a synapse.
17. List some of the processes controlled by the cerebrum.
18. Describe the functions and structure of the outer, middle, and inner ear.
19. Name and describe the three major parts of a neuron.
20. Describe how an impulse travels down a neuron.

Chapter 33
Reproductive System

ALL ORGANISMS PRODUCE more individuals of their kind through the process of reproduction. Some organisms simply split in two, while others have developed complicated sexual reproduction systems.

Sexual reproduction involves the joining of two **gametes,** or sex cells. The egg is the gamete produced by the female, and the sperm is the gamete produced by the male. During the process of sexual reproduction, a new individual is formed. The **gonads** are the reproductive organs that produce gametes. The gonads are also endocrine glands that produce hormones that regulate the development of certain physical characteristics and the production of gametes.

The male and female reproductive systems produce gametes that carry important instructions in the form of DNA. These two gametes develop into an individual with genetic information from both parents. However, because the DNA came from both parents, offspring contain genetic information that is different from that of either parent. The differences that result from sexual reproduction are important factors in the evolutionary changes that occur in any sexually reproducing species.

THE MALE REPRODUCTIVE SYSTEM

The male gonads are the **testes.** The male has two testes. These organs are responsible for producing the hormones that determine male secondary sex characteristics such as facial hair, a deeper voice, and larger muscles. The testes are also responsible for producing the male gametes, or **sperm.** The testes are located in a saclike structure called the **scrotum,** which hangs outside the body cavity. The temperature in the scrotum is a few degrees lower than normal body temperature. This slightly lower temperature is necessary for the production of sperm.

Sperm are actually made in small coiled tubes inside the testes. These tubes are called **seminiferous tubules.** The seminiferous tubules connect to larger tubes located on top of the testes. These tubes are called the **epididymus.** They store the sperm after they are produced in the testes. Each epididymus is connected to another tube called the **vas deferens.** The two vas deferens, one from each testis, connect to a single tube, the **urethra,** which passes through the penis and opens to the outside of the body. The **prostate, seminal vesicles,** and **Cowper's gland** connect to the urethra and add secretions, called **seminal fluid,** to the sperm. Seminal fluid provides nourishment for the sperm and helps the sperm become more motile. The fluid also helps neutralize the acidity in the urethra caused by urine, which could kill the sperm. Together, the seminal fluid and the sperm are called semen.

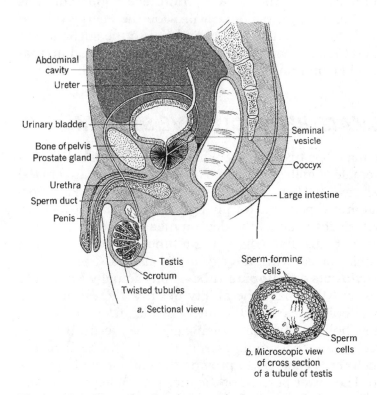

Figure 33-1. Reproductive system of a human male.

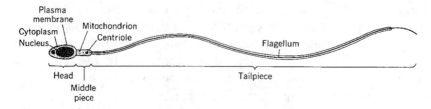

Figure 33-2. A sperm cell.

SPERM

Sperm are formed in the testes through a process called **spermatogenesis.** Sperm carry the male genetic information. Each sperm is made up of three distinct parts: the head, the neck or midpiece, and the tail. The **head** contains a haploid nucleus. This means that it carries one half the number of chromosomes contained in a normal body cell. In humans the haploid number is 23. The **neck** or **midpiece** region contains many mitochondria. Mitochondria provide the energy needed by the sperm to move. The mitochondria use the sugar in the seminal fluid to make packets of ATP. The **tail** is a flagellum that creates wavelike motions to propel the sperm forward.

THE FEMALE REPRODUCTIVE SYSTEM

The female has two ovaries, which are the organs that produce the female gametes, or **eggs.** The **ovaries** are located in the abdominal cavity. They produce not only eggs but also hormones that are responsible for giving the female her secondary sex characteristics such as the development of the breasts and reproductive organs. Each ovary has a tube located near it that is responsible for carrying the egg to the uterus. These tubes, called the **oviducts** or **fallopian tubes,** don't actually connect to the ovaries, but they have the ability to draw the egg into the tube. The fallopian tubes are the site of fertilization. Both tubes empty into a hollow muscular organ called the uterus. If an egg is fertilized, it attaches, or implants, itself to the wall of the uterus. The uterus expands to provide the embryo with a place to develop. The lower portion of the uterus is called the **cervix.** The cervix opens into the **vagina,** which opens to the outside of

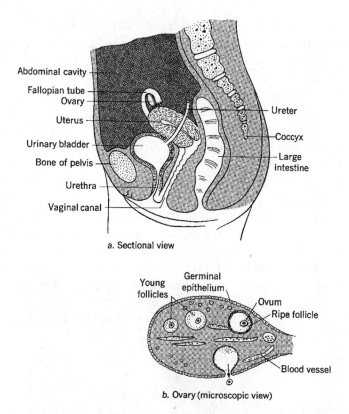

a. Sectional view

b. Ovary (microscopic view)

Figure 33-3. Reproductive system of a human female.

the body. The vagina serves as the point of deposit for the sperm and as the birth canal for the baby.

THE MENSTRUAL CYCLE

Each month, changes occur in the female reproductive system. These changes are controlled by hormone levels in the blood. This series of changes is called the **menstrual cycle.** During the menstrual cycle, an egg is released, and the uterus prepares to receive the egg. If the egg is not fertilized, it passes out of the female's body along with the lining of the uterus. The menstrual cycle lasts about 28 days and has four distinct stages. The **follicle stage** begins when the pituitary gland starts to secrete **follicle stimulating hormone** (FSH). This hormone causes a layer of cells to form around an egg. This layer of cells is called the **follicle.** As the follicle grows, it secretes the hormone **estrogen,** which causes the lining of the uterus to become thicker, making it more suitable to receive a fertilized egg. The lining of the uterus is called the endometrium.

When estrogen in the blood reaches a certain level, the pituitary begins to secrete **luteinizing hormone** (LH). This hormone causes the follicle to rupture, releasing the egg. This process is called **ovulation.** Ovulation is the second stage of the menstrual cycle.

The third stage is the **corpus luteum** stage. After ovulation has occurred, the follicle becomes a yellow structure called the corpus luteum. The hormone **progesterone** is secreted by the corpus luteum. This hormone causes the lining of the uterus to continue to build up. If the egg is fertilized, progesterone continues to be secreted. If fertilization does not occur, the corpus luteum breaks down and the level of progesterone drops. The lack of progesterone causes the lining of the uterus to break down, and then pass out of the body through the vagina. The

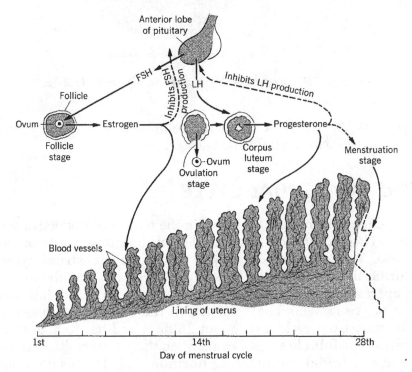

Figure 33-4. The menstrual cycle.

discharge of the lining of the uterus and the unfertilized egg is the last stage of the cycle and is called **menstruation.** Menstruation occurs in most women until they reach their mid-40s. At this time, they stop releasing eggs. This affects the production of hormones that control the woman's cycle. **Menopause** is the term used when a woman stops having a menstrual cycle.

THE EGG

The female gamete is an egg, or **ovum**. Eggs develop in the ovaries. The production of eggs is called **oogenesis.** A woman is born with several hundred thousand immature eggs in her ovaries. During a woman's fertile years, only a few hundred eggs actually mature and pass through her reproductive tract. Eggs are much larger than the sperm produced by males.

DEVELOPMENT OF THE EGG AFTER FERTILIZATION

The uniting of an egg and sperm is called fertilization. The fertilized egg is called a **zygote.** During the first few days of its development, the zygote undergoes many mitotic divisions. This

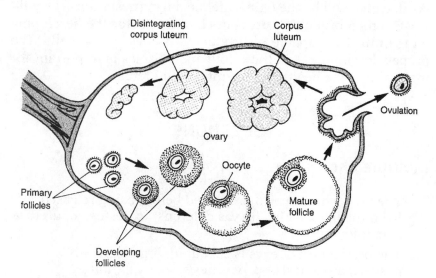

Figure 33-5. Ovulation.

series of divisions is called **cleavage.** Although the number of cells increases, the size of the new cells gets smaller. After several divisions, a cluster of cells called the **morula** forms. As cell division continues, a hollow ball of cells forms. This stage is called the **blastula.** The hollow area in the center of the ball of cells is called a **blastocyst.** Eventually an opening, called the **blastopore,** forms in one area of the sphere of cells. This opening is actually the folding in of the outer cell layer. As the fold increases in size, three cell layers begin to form. These three layers are called the endoderm, mesoderm, and ectoderm. The **endoderm** develops into the digestive and respiratory systems. The **mesoderm** eventually forms the skeleton, muscles, excretory system, and circulatory system. The **ectoderm** forms the nervous system and skin.

The period of development of a fertilized egg in the uterus is called **gestation.** The fertilized egg is called an **embryo** during its first two months of development. After that, it is called a **fetus.**

Four membranes develop around the embryo to provide protection and nourishment. The **chorion** provides an area for the exchange of food, oxygen, and wastes between the embryo and mother. The **placenta** is formed from the chorion and provides a point of contact with the mother's blood supply. Another membrane, the **amnion**, forms a protective fluid-filled sac around the fetus. The **allantois** membrane helps form the **umbilical cord**, which will be the fetus's lifeline during gestation. The **yolk sac** is a membrane that produces blood cells for the developing fetus until it is capable of producing its own blood cells. The proper development of these four membranes is important for maintaining the health of the embryo and fetus.

QUESTIONS

Multiple Choice

1. The area where sperm are stored before being released from the body is the   *a.* vas deferens   *b.* testes   *c.* urethra   *d.* epididymus.
2. The production of eggs is called   *a.* organogenesis   *b.* spermatogenesis   *c.* oogenesis   *d.* ovulation.

3. Which of the following hormones, released from the pituitary gland, is responsible for the growth of a layer of cells around the egg? *a.* FSH *b.* LH *c.* estrogen *d.* progesterone

4. During the development of the zygote, rapid cell division occurs. This cell division is called *a.* gestation *b.* ovulation *c.* cleavage *d.* fertilization.

5. The egg and the sperm are both *a.* diploid cells *b.* gonads *c.* zygotes *d.* haploid cells.

Matching

6. fertilization
7. menstruation
8. ovulation
9. oviducts
10. gonads

a. release of the egg from the follicle
b. joining of the egg and the sperm
c. ovaries and testes
d. the fallopian tubes
e. shedding of the lining of the uterus

Fill In

11. The _____ is the hollow area in the center of the blastula stage of development.

12. The _____ _____ are coiled tubes found in the testes where sperm are produced.

13. The hormone _____ is secreted by the corpus luteum.

14. The formation of the sperm in the testes is called _____.

15. Implantation of the egg occurs in the _____.

Free Response

16. Describe the structures and functions of the organs of the female reproductive system.

17. Describe the functions of the hormones involved in the female menstrual cycle.

18. Describe the structures and functions of the organs of the male reproductive system.

19. How does sexual reproduction contribute to the variation that occurs in humans?

Chapter 34
Disease and Immunity

EVERY DAY OUR bodies are exposed to millions of harmful organisms and substances. Our bodies have many barriers to protect us from these foreign invaders. The skin and mucous membranes serve as a first line of defense. If the invaders get through the skin and membranes, the immune system rushes into action in an effort to destroy the harmful visitors.

THE CAUSES OF DISEASE

If a foreign invader enters the body and is allowed to multiply, disease is the result. **Diseases** cause the body to malfunction, and we become sick. If the disease is **infectious,** it can be spread from host to host. **Louis Pasteur** was the first scientist

Figure 34-1. Louis Pasteur.

to determine that microorganisms cause diseases. In the late 1800s, **Robert Koch** developed a method to determine which organism causes a particular disease. His procedure became known as **Koch's postulates.** He first found the microorganism in the diseased plant or animal. He removed it from the host and cultured it. After he had a pure culture, he took some of the organisms and injected them into a healthy host. If the organisms produced the symptoms of the disease and were found in the infected hosts, they were determined to be the cause of the disease.

Today, we know that many diseases are caused by bacteria. Some disease-causing bacteria produce **toxins,** chemicals that are poisonous to cells. Tetanus is a disease caused by toxins. Some bacteria actually enter tissues and cells, where they cause the damage themselves. Tuberculosis is a disease caused by this type of bacteria. Viruses also cause disease. Viruses are parasites in living cells. Colds and flus are both viral diseases.

THE BODY'S REACTIONS

The job of the **immune system** is to protect our bodies from any disease-causing organisms that gain entry. This system is able to tell the difference between disease-causing substances and harmless substances. If the substance is harmful, an **immune response** begins to destroy the invader. The substance that causes an immune response is an **antigen,** or foreign protein.

Phagocytes, a type of white blood cell, play a major role in protecting our body. These blood cells surround, attack, and destroy the unwanted invader. **Lymphocytes** also help in the protection of the body. They are another type of white blood cell and are found in two forms, T cells and B cells. Both the T cells and B cells are made in the bone marrow.

T cells recognize and attack antigens. **Helper T cells** are able to link to antigens. This linkage causes the activation of killer T cells and B cells. **Killer T cells** cause the death of any diseased cells. **B cells** produce antibodies to destroy disease-causing substances. **Antibodies** are proteins that are able to fight infection. They are produced in large amounts during an infection.

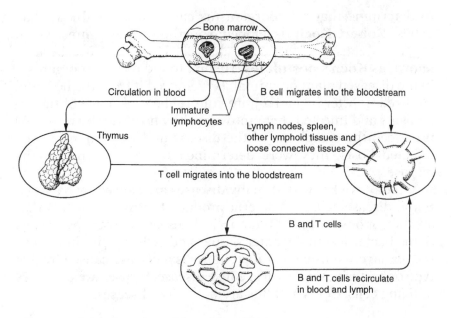

Figure 34-2. Origins of B cells and T cells.

Some of the T cells and B cells remain after the invading organisms have been destroyed. They are called **memory cells** and provide the body with the **immunity** needed to resist the invaders in the future. This is called **active immunity.** Humans also have **passive immunity.** This occurs when antibodies are introduced into the body. When a baby receives antibodies from its mother, it acquires a type of natural immunity.

Viruses that enter the body are attacked by a protein called **interferon.** This protein prevents viruses from reproducing. It seems to be effective against many different viruses. Biotechnologists have discovered ways to produce human interferon.

MEDICINES AND DISEASE

Sometimes our bodies need help in preventing disease and sickness. The use of chemicals to treat disease is called **chemotherapy**. Many antibiotics have been developed to fight bacterial diseases. They are able to inhibit the growth of, or destroy, microorganisms. **Alexander Fleming** discovered antibiotics in

1929. He observed that the mold *Penicillium* seemed to stop the growth of bacteria. From this mold, the antibiotic penicillin was produced. Penicillin is probably the best known antibiotic. Today, antibiotics are produced both synthetically and in living organisms.

Vaccines also are used to prevent diseases. **Vaccines** are usually made up of weakened disease organisms. Even though they won't give you the disease, the vaccines will cause your body to produce antibodies and memory cells. This usually provides immunity for a long time. The first vaccine was developed by **Edward Jenner** in 1798. His vaccine for smallpox has saved the lives of thousands of people. The effectiveness of vaccines against smallpox has eliminated this disease from the human population.

QUESTIONS

Multiple Choice

1. A substance that causes an immune response is called a(n)
   a. antibiotic   *b*. antigen   *c*. B cell   *d*. T cell.
2. Viruses that enter the body are attacked by
   a. antibiotics   *b*. antibodies   *c*. antigens   *d*. interferon.
3. Memory cells provide the body with   *a*. immunity
   b. interferon   *c*. antigens   *d*. toxins.
4. The use of chemicals to treat disease is called   *a*. vaccination   *b*. an immune response   *c*. chemotherapy
   d. toxicology.
5. Proteins that are able to fight infection are called
   a. antigens   *b*. antibiotics   *c*. antibodies   *d*. toxins.

Matching

6. B Cells   *a*. poisonous to cells
7. helper T cells   *b*. produce antibodies
8. killer T cells   *c*. provide immunity
9. vaccines   *d*. link to antigens
10. toxins   *e*. kill diseased cells

Fill In

11. T cells recognize and attack _____.
12. _____ cause the body to malfunction, and we become sick.
13. The job of the _____ system is to protect our bodies from disease.
14. The _____ and mucous membranes serve as our first line of defense against harmful invaders.
15. The best-known antibiotic is _____.

Free Response

16. Describe how Koch would determine a disease-causing agent.
17. Compare the functions of B and T cells.
18. Describe the different ways immunity can be gained.
19. How does the human body protect itself from attack by viruses?
20. How do vaccines protect the body?

Portfolio

Many physicians are concerned that certain types of bacteria are becoming resistant to antibiotics. Interview a physician about the dangers posed by drug-resistant bacteria. You might like to write an article for your school newspaper about this problem.

UNIT TEN
BEHAVIOR OF LIVING THINGS

Chapter *35*

Drugs, Tobacco, and Alcohol

DRUGS ARE SUBSTANCES that cause changes in the way the body functions. Physicians often administer drugs to improve a person's health. For example, a bacterial throat infection may be treated with antibiotics. Antibiotics help the body fight the bacterial infection and return it to a normal healthy state. Antibiotics and other medicines are known as **prescription drugs.** These drugs are often very powerful and may pose a danger if they are misused. Some have unpleasant side effects, or they may interact with other drugs and substances and be harmful. The sale of prescription drugs is regulated by law to make sure they do not harm an individual. Prescription drugs must be taken according to a physician's directions.

Other medicinal drugs can be purchased without a prescription. These are called **over-the-counter drugs.** These drugs are safe for most individuals if the instructions are followed carefully. Aspirin is a good example of an over-the-counter drug. Over-the-counter drugs also may be dangerous if they are misused. They may interact in a harmful way with other drugs or foods.

THE EFFECTS OF DRUGS ON THE BODY

Drugs can be categorized by how they affect the way the body functions. **Stimulants** are drugs that increase the activities of the nervous system. As a result, heart rate and blood pressure increase. Caffeine, amphetamines, and cocaine are classified as stimulants. Caffeine is found in tea, soft drinks, and coffee. Physicians advise that you limit your daily intake of caffeine. Excess amounts of caffeine may cause rapid heartbeat and other unpleasant reactions. Amphetamines, the so-called diet pills, are also dangerous to take. Cocaine is a very dangerous drug. Cocaine may cause serious psychological disturbances. In rare cases, cocaine use may prove fatal.

Another group of drugs are the **depressants.** Depressants decrease the activity of the nervous system. They are sometimes called sedatives. Tranquilizers and barbiturates act as depressants. Valium is an example of a prescription tranquilizer. **Narcotics,** such as heroin, codeine, and morphine, are depressants

Table 35-1. Dangerous Drugs and Their Effects

Drug Type	Examples	Effects
Depressants	Alcohol, valium, barbiturates (i.e., nembutal, seconal)	Slow central nervous system; modify behavior (loss of anxiety); induce sleep; slurred speech; anesthesia; delirium; death from overdose
Stimulants	Cocaine, crack, amphetamines, caffeine, nicotine	Give "highs"—increased alertness to sound and sight; reduce fatigue; elevate mood. Lead to depression; increased heart rate; loss of appetite; death from overdose
Narcotics	Heroin, opium, codeine	Sedate body; reduce pain; some are extremely addictive; death from overdose
Hallucinogens	LSD, PCP, marijuana, hashish	Induce hallucinations and psychoses; alter perceptions; poor motor control; confusion

and are very addictive. The drugs mentioned above have very powerful effects on the body. They may cause drowsiness and decrease the body's ability to coordinate its movements. Many deaths have resulted from the misuse of depressants.

There is also a group of drugs that distorts many of the sensory messages. These drugs are called **hallucinogens.** They directly affect the central nervous system. **LSD** and **PCP** (angel dust) are both hallucinogens. The effects of LSD and PCP vary from person to person. LSD has been known to cause severe distortions of reality (bad "trips") in some people. The distortions can become frighteningly real, and very little can be done to relieve a person's suffering until the effects of the drug wear off. PCP is a very dangerous drug that distorts reality in the user and produces violent behavior in some individuals. **Marijuana,** another hallucinogen, is probably the most widely used illegal drug. Long-term marijuana use can be dangerous. It contains harmful substances, including carcinogens, that are inhaled along with the smoke.

After using drugs repeatedly, the body may form a physical and/or psychological dependence on the drug. The body needs more and more of the drug until the user forms an **addiction** to the drug. If the user tries to stop using the drug, he or she may suffer **withdrawal** symptoms, including depression, psychological disturbances, and convulsions. Think about this. Even the use of prescription drugs may harm the body if they are taken improperly. Prescription drugs are made in factories under extremely clean and precise conditions. The doses are exact in each pill, capsule, or liquid. Illegal drugs are made in labs that may not adhere to rigorous rules of cleanliness. Illegal drugs are often contaminated with substances that may in themselves be fatal. They are made by people who do not make sure that there is a standard dose delivered. Taking illegal drugs, and taking prescription drugs improperly, is a very dangerous practice.

TOBACCO

Tobacco products are made from the dried and cured leaves of the tobacco plant. Tobacco may be smoked, chewed, or inhaled in a powder known as snuff. Tobacco contains many different chemicals. Nicotine is the main drug in tobacco. Nicotine is a very potent chemical. In concentrated form, it is a poison. In the small amounts found in tobacco products, nicotine acts as a stimulant. Nicotine increases the activities of the nervous system. It is also very addictive. People who are addicted to tobacco products experience withdrawal symptoms when they stop using them.

Tobacco also contains tars. The chemicals in tar injure the tiny hairs, or cilia, that line the breathing passages. These hairs are one of the body's main defenses against inhaling harmful substances into the lungs. In many studies, tars have been proved to cause cancers. Tars in tobacco smoke have been shown to cause lung cancer. Tars in chewing tobacco have been shown to cause cancers of the tissues in the mouth. Carbon monoxide, a chemical produced when tobacco burns, is a poisonous gas. Carbon monoxide decreases the body's abilities to transport oxygen to its cells.

Tobacco use also affects other body systems. By increasing blood pressure, the chemicals in tobacco increase the stress on the circulatory system, including the heart. Smoking tobacco increases the risk of death from heart disease. Tobacco smoke also poses a danger to nonsmokers who inhale the smoke of others. Many states and municipalities prohibit smoking in buildings and other indoor spaces. Inhaling smoke is dangerous to young children. Since their lungs are smaller than those of adults, children are more easily and seriously affected by inhaling cigarette smoke.

Smoking shortens a person's life. Smoking also damages a person's health. For good health and an increased life span, it is best never to smoke. People who smoke are advised by physicians and health professionals to stop smoking.

ALCOHOL

One of the most used and abused drugs in our society is **alcohol.** There are many types of alcohol, but it is **ethyl alcohol** (C_2H_5OH) that is found in alcoholic beverages. Alcohol is produced by a process called fermentation. **Fermentation** is the anaerobic respiration of yeast, which uses the sugar found in grains and fruits as a food source. During fermentation, the yeasts produce alcohol as a waste product.

Alcohol is a depressant that is absorbed directly into the bloodstream from the stomach and intestines. The amount of alcohol present in different beverages varies. The amount of alcohol in the bloodstream is called the **blood alcohol concentration** (BAC). While it travels through the bloodstream, alcohol passes through the brain. This can produce poor judgment, a lack of coordination, and a slowing of reaction time. That is why it is so dangerous to drink and drive! An extremely high level of alcohol in the blood can cause the respiratory center of the brain to stop sending signals to the body and result in death. The brain is not the only organ affected by alcohol. The function of the liver and kidneys is also impaired as a result of alcohol in the blood.

Gastritis is an irritation of the stomach that can result from excessive use of alcohol. Sometimes, in the liver of a heavy alcohol user, healthy tissue is replaced by scar tissue. This condition is called cirrhosis. **Cirrhosis** can damage the liver to the extent that the person dies. Even a fetus can be harmed by alcohol. Women who drink during pregnancy may give birth to children who have **fetal alcohol syndrome** (FAS). These babies may suffer from short-term and long-term health problems, including learning difficulties and heart problems.

Just as they do with other drugs, some people become dependent on alcohol. This disease is called **alcoholism.** Large

Table 35-2. The Effects of Alcohol on the Brain

BAC (Blood alcohol concentration)	Behavior
0.05%	Judgment suffers, inhibitions reduced
0.1%	Reaction time reduced, difficulty driving and walking
0.2%	Abnormal behavior, sadness, weeping
0.3%	Vision and hearing impaired
0.45%	Unconscious
0.65%	Death

amounts of alcohol, over long periods of time, harm the body and shorten life expectancy. The effects of alcohol on the body may become stronger or more dangerous when alcohol is used along with other drugs.

Drugs are intended to help maintain a healthy body. When they are misused, they do the opposite. Abuse of drugs can weaken body tissues, and even be fatal.

QUESTIONS

Multiple Choice

1. Which groups of drugs increases the activity of the nervous system? *a.* depressants *b.* stimulants *c.* narcotics *d.* tranquilizers

2. The type of alcohol found in alcoholic beverages is *a.* ethyl alcohol *b.* butanol *c.* isopropyl *d.* methanol.

3. Which of the following is not a stimulant? *a.* barbiturate *b.* cocaine *c.* amphetamine *d.* caffeine

4. Alcohol is produced by a process called *a.* photosynthesis *b.* aerobic respiration *c.* Kreb's cycle *d.* fermentation.

5. Cirrhosis is a damaged condition of the *a.* heart *b.* lungs *c.* brain *d.* liver.

Matching

6. addiction   *a.* addictive depressants

7. narcotics   *b.* amount of alcohol in blood

8. alcoholism   *c.* dependency on alcohol

9. BAC   *d.* sedatives

10. depressants   *e.* dependency on a drug

Fill In

11. _____ are substances that cause the body to function differently than normal.

12. Fermentation is a type of_____ respiration.

13. _____ is an irritation of the stomach sometimes caused from too much alcohol.
14. If a drug abuser tries to stop using the drug, he or she may suffer from _____.
15. PCP and LSD are both _____.

Free Response

16. What are some of the effects of alcohol on the body?
17. How are depressants different from stimulants?
18. Can prescription drugs be harmful? If so, how?
19. How does alcohol affect the liver?
20. How could drinking alcohol be fatal?

Portfolio

Arrange for a drug counselor to speak with your class. These people are often former drug abusers, and they can speak meaningfully about their experiences. You might like to write a short summary for the school newspaper about what was discussed.

Chapter 36

Plant Tropisms

IF YOU WATCH a field of sunflowers for a day, you will observe that the plants follow the movement of the sun. It's actually quite remarkable. In the morning, the sunflowers turn toward the east; in the afternoon, they face the sun setting in the west.

Plants respond to many different stimuli. For example, water and gravity cause plants to respond in different ways. A **tropism** is the way a plant responds to an environmental stimulus, such as the sun. Tropisms involve plant growth, so they are not reversible. The response from a plant is considered positive when the plant grows toward a particular stimulus. The response is considered negative when the plant grows away from a particular stimulus.

KINDS OF TROPISMS

Tropisms are put into categories according to the particular kind of stimulus that initiates the response. For example, a plant's response to light is called **phototropism.** (The prefix *photo-* means "light.") This is usually a positive tropism in which a plant grows toward the light. Plants that grow on a windowsill must be regularly rotated to promote even growth. Charles Darwin first described the process of phototropism. He and his son observed that oat seedlings tended to bend toward a source of light as they grew. They concluded that something was being

Figure 36-1. Phototropism.

produced at the tip of the oat seedling and then sent down the stem. This substance caused the bending of the plant toward the light. Today we know that the substance the Darwins described was a type of plant hormone called auxin.

AUXIN

Auxin is a naturally occurring hormone that regulates plant growth. In response to light, auxin moves from the bright side of the plant, the side facing the sun, to the dark side of the plant, the side that does not receive the direct rays of the sun. Auxin collects in the cells on the dark side of the stem. These cells are stimulated to grow longer, or elongate. Because the cells on the dark side of the stem grow longer than the cells on the sunlit side of the stem, the stem actually bends toward the light.

A plant's response to water is called **hydrotropism.** (The prefix *hydro-* means "water.") Roots grow toward water. This is a positive tropism. Occasionally, tree roots searching for water actually grow into, and clog, water pipes.

Geotropism or **gravitropism** is a plant's growth response to gravity. Different parts of a plant respond differently to gravity. Therefore, a plant can show both positive and negative responses to gravity. An example of positive geotropism is roots

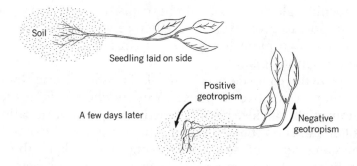

Figure 36-2. Geotropism.

growing toward a source of gravity—down toward the Earth's center. By growing down into the soil, the roots can take in important nutrients and water that the plant needs to live. An example of negative geotropism is a plant's stem growing away from a source of gravity—up toward the sun. Auxin regulates positive and negative geotropism, just as it controls phototropism. Wherever it accumulates in a plant, auxin causes the plant's cells to grow longer.

EVIDENCE THAT SHOWS PLANT RESPONSES TO GRAVITY

A variety of simple experiments can be performed to illustrate both types of geotropism. Seeds seem to grow properly, no matter how they are planted. The shoot, which is the plant stem, will show a negative geotropism, while the roots will show a positive geotropism. Scientists think that **amyloplasts**, plant cell organelles that contain starch, may influence geotropism. Starch grains within the amyloplasts seem to collect where gravity pulls on them. This movement may cause auxins to be redistributed in the plant, resulting in cell elongation.

PLANTS RESPOND TO TOUCH AND CHEMICALS

Thigmotropism is a plant's response to touch. This tropism causes some plants to curl around objects. For example, grapes and morning glories show thigmotropism. These vines, which need support to grow well, wrap themselves around any object they touch. Two plant hormones, auxin and ethylene, seem to control thigmotropism.

A growth response to chemicals is called **chemotropism.** Roots show a positive response to many chemicals in the soil. Chemotropism can also be observed during plant fertilization. After pollen lands on the female reproductive structure of a flower, it grows a pollen tube down to the ovule. This pollen tube is necessary in order for the sperm and egg cell to unite. Chemicals in the female structure are responsible for guiding the growth of the pollen tube. This is a good example of chemotropism.

Nastic Movements

Nastic movements are plant movements that are independent of the direction of the stimulus. For example, some flowers open during the day and close at night. Other flowers open in the evening and close during the day. Another nastic movement is the opening and closing of the Venus's-flytrap's leaf traps. Nastic movements do not involve cellular growth, so they are reversible.

All living things detect and respond to the environment. As you can see, even without a nervous system, plant behavior is very complex. It involves external stimuli, such as light and water, and internal chemical messengers.

QUESTIONS

Multiple Choice

1. A plant's response to water is called   *a.* thigmotropism   *b.* geotropism   *c.* phototropism   *d.* hydrotropism.
2. Which of the following scientists first described phototropism?   *a.* Aristotle   *b.* Darwin   *c.* Watson   *d.* Linnaeus
3. Which of the following hormones regulates phototropism?   *a.* auxin   *b.* ethylene   *c.* gibberellin   *d.* cytokinin
4. Tropisms are categorized by their particular kind of   *a.* response   *b.* movement   *c.* stimulus   *d.* hormone.

5. When morning glories curl around another plant, it is an example of   *a*. thigmotropism   *b*. nastic movement   *c*. geotropism   *d*. phototropism.

Matching

6. phototropism
7. nastic movement
8. geotropism
9. chemotropism
10. hydrotropism

a. independent of the direction of the stimulus

b. response to light

c. response to gravity

d. response to water

e. response to chemicals

Fill In

11. When a plant grows away from a stimulus, it is called a _____ response.
12. When auxin collects in cells, the cells tend to _____.
13. _____ is the way a plant grows in response to a stimulus.
14. Charles Darwin was the first scientist to describe _____.
15. Hydrotropism can be seen when _____ grow toward water.

Free Response

16. How are nastic movements different from thigmotropism?
17. Explain how auxin causes a plant stem to turn toward light.
18. Explain how amyloplasts influence geotropism.
19. How are phototropism and geotropism similar?

Portfolio

Design an experiment that shows the effects of gravity on the growth of a seedling's roots. Radish seeds sprout quickly and work quite well in this type of experiment. Make a drawing or take photographs that show the result of your experiment. Use these illustrations in your written laboratory report.

Chapter 37
Biological Rhythms

DID YOU EVER wonder why owls hunt at night and eagles during the day? Why some plants fold up their leaves at night? Many biological processes undergo regular changes that are related to time. Often these changes seem to be repeated in a regular manner and are referred to as **rhythmic behavior.** Examples of rhythmic behavior are feeding during the day and resting at night.

THE BIOLOGICAL CLOCK

Organisms that are active during the day are called **diurnal.** Humans are primarily diurnal organisms, although some humans are active at night—either by their choice of work or by their choice of recreation. Often, people who work at night undergo a period of discomfort as their body adjusts to being active at night. Other organisms are always active at night and rest during the day. These organisms are called **nocturnal.** Owls are nocturnal, hunting for small mammals that are active at night.

One of the first written accounts of biological rhythms was an early naturalist's observations of legumes, plants in the bean and pea family. He described the pattern of leaf movements during an entire day. Today, when his legume experiments are duplicated in a laboratory, we see the same results. Even when the plants are deprived of external cues about the time of day or night, the leaf movements continue to remain synchronized with those of the same type of plants in nature.

Discoveries like these demonstrated to scientists that organisms have some way of "keeping" time. This mechanism of keeping track of time is referred to as a **biological clock.** Biological clocks have the ability to reset themselves. However, to do this, they must be affected by an external stimulus. These external stimuli are called **zeitgebers.** The most common zeitgeber is light. When organisms are deprived of zeitgebers, their original rhythms continue, but their cycle may become longer.

PLANT RHYTHMS

In plants, the blue-green pigment **phytochrome** seems to be an important part of the biological clock. The pigment is found in two forms, Pr (phytochrome red) and Pfr (phytochrome far-red). Pr has the ability to absorb red light during daylight hours. When it absorbs red light, the Pr is converted into Pfr. At night, the Pfr is converted back into Pr. The reversible pigment phytochrome may be the clock that allows plants to determine the length of days.

Plants also respond to changes in the length of daylight. **Photoperiodism** is a plant's response to varying periods of light. The flowering of many plants is affected by the length of daylight. Plants can be divided into three groups based on their response to day length: short-day plants, long-day plants, and day-neutral plants. **Short-day plants** usually flower in autumn

Figure 37-1. Poinsettia, a short-day plant.

when the period of daylight grows shorter. **Long-day plants** bloom in spring and summer when the days are long. **Day-neutral plants** don't seem to respond to any particular day length and can flower at any time of the year.

ANIMAL RHYTHMS

In the last 25 years, rhythms have been described in all major groups of plants and animals. These rhythms are not learned. Animals raised from birth in a controlled laboratory setting exhibit certain rhythmic behaviors when subjected to an appropriate stimulus.

It was thought that the endocrine system and the nervous system helped regulate the biological clock in animals. However, rhythms have even been discovered in unicellular organisms that lack complex hormonal and nervous systems. This evidence seems to point to the individual cell as the site of the biological clock. Researchers have tried to locate a biological clock within the cell. The most logical place to search is the nucleus. However, even after scientists remove the nucleus from cells, basic rhythms are still present.

Biological clocks that operate on a monthly or yearly basis exist, but most studies have been conducted on biological clocks that operate on a 24-hour cycle. Because the cycles may be slightly longer or shorter than 24 hours, these rhythms are called **circadian rhythms**. (*Circa* means "about," and *diem* means "day.") Circadian rhythms can even be found in many of the microscopic animals that live in the ocean.

Some patterns in plants and animals occur in yearly cycles. These are called **annual rhythms.** Many reproductive cycles are annual. For example, the young of many animals are born in warm weather. This allows for a period of growth and increased strength before the cold winter. Hibernation and estivation are also annual rhythms. Both of these conditions allow an organism to conserve energy. **Hibernation** is a slowing down of physiological activity during the winter. Bears and other animals hibernate. **Estivation** is reduced activity during the summer. Some frogs estivate in warm weather when their ponds dry up.

Seasonal changes can also create rhythms. Many animals migrate because of changing seasons. These animals have to be at an appropriate feeding area when food is available. The availability of food is usually due to seasonal climate changes. The movements of many herding mammals in Africa and many birds in North America are caused by **migratory rhythms.**

Tidal rhythms, or **lunar rhythms,** are caused by the position of the sun and moon. Many animals that live near the shoreline respond to the rhythmic rise and fall of the ocean. Crabs have been a favorite animal subject in the study of tidal rhythms. The fiddler crab comes out of its burrow at low tide to feed. At high tide, it retreats back into its burrow.

Many behavioral and physiological processes occur in rhythmic patterns that take place over days, months, and even years. Since these cycles are controlled by biological clocks in individual cells, there are many clocks in each living organism. These biological clocks permit an organism to compare its internal rhythms with the constantly changing environmental rhythms around it. Scientists believe there is a master clock that is responsible for keeping the other clocks synchronized. The study of biological rhythms and clocks will be an important field of research in the future, as scientists try to understand more about the timekeeping mechanism, the ability to reset the clock, and the communication mechanism between the clock and the resulting behavior.

QUESTIONS

Multiple Choice

1. The slowing down of physiological activity during the winter is   *a.* estivation   *b.* hibernation   *c.* tidal rhythms   *d.* lunar rhythms.

2. A plant's response to different amounts of light is   *a.* photosynthesis   *b.* estivation   *c.* phytochrome   *d.* photoperiodism.

3. Biological clocks can be reset by external stimuli called   *a.* zeitgebers   *b.* phytochromes   *c.* rhythms   *d.* circadians.

4. Animals that are active at night and that rest during the
   day are called   *a.* diurnal   *b.* tidal   *c.* nocturnal
   d. zeitgebers.
5. Fiddler crabs exhibit   *a.* circadian rhythms   *b.* tidal
   rhythms   *c.* migratory rhythms   *d.* photoperiodism.

Matching

6. tidal rhythms   *a.* about a day
7. circadian rhythms   *b.* caused by sun and moon
8. annual rhythms   *c.* due to seasonal changes
9. migratory rhythms   *d.* plant pigment
10. phytochrome   *e.* yearly cycle

Fill In

11. Organisms that are active during the day are called _____.
12. Phytochrome is found in two forms, phytochrome red and
    phytochrome _____.
13. The _____ and _____ systems were the first systems to be
    suspected to have a relationship with the biological clock.
14. The most common zeitgeber is _____.
15. Hibernation and estivation are both _____ rhythms.

Free Response

16. Compare estivation and hibernation and give examples of
    animals that have these cycles.
17. What does the biological clock do for organisms?
18. How do plants "tell time"?
19. What is the advantage of some animals being diurnal and
    others being nocturnal?
20. What is the importance of migration to some animals?

Portfolio

Keep a list of your activities for several days. Do you find that
you repeat certain behaviors at certain times?

Chapter 38

Behavior: Innate vs. Learned

IF YOU HAVE ever trained a puppy, you know that you have to change some of the ways a puppy acts. For example, in order for a dog to become a good pet, you have to train it to walk on a leash, obey commands, not to chew furniture, not to bite, and other important skills. In effect, you have to change its behavior.

Behavior is anything an organism does or how it acts. An organism's behavior helps it survive and be successful. The behavior is affected by the organism's environment and by its genetic program.

Scientists have determined that most living things exhibit two types of behavior: innate behavior and learned behavior. **Innate behavior** is inborn and works perfectly the first time it happens. **Learned behavior** is acquired during the life of the organism and may take a long time to perfect.

INNATE BEHAVIOR

Innate behavior is determined by an organism's genetic makeup, as are eye color and some other physical features. Humans are born with the ability to suckle. They even show this behavior in the uterus. Thus, they are able to nurse as soon as they are born. Even though innate behavior is not learned, it can change through the many experiences an organism has during

Figure 38-1. A bird's ability to build a nest is instinctive.

its lifetime. Many very complex innate behaviors, sometimes called **instincts,** are found in animals. Nest building in birds and the spinning of webs by spiders are both complex instinctive behaviors.

OTHER TYPES OF INBORN BEHAVIOR

Another example of an inborn behavior is a taxis. A **taxis** is a response made by a whole organism to an environmental stimulus. **Geotaxis** is a response to gravity. **Phototaxis** is movement in response to light. A response to chemical substances is called **chemotaxis.** If an organism moves toward a stimulus, this is a **positive response.** If the organism moves away from the stimulus, it is a **negative response.** An example of a negative response is a cockroach's movement away from light. This is an example of a negative phototaxis.

A reflex is another kind of innate behavior. A **reflex** is a response that does not involve complex thought processes. It is a rapid response that is beneficial to the organism. Sneezing and blinking are both reflexes. Reflexes occur in a certain pattern called a **reflex arc.** A reflex arc involves **receptors** that receive the stimulus; **sensory neurons** that transmit the impulse to the spinal cord; **associative neurons** or **interneurons** that make neural connections in the spinal cord; and **motor neurons** that carry the impulse from the spinal cord to an **effector,** usually a gland or muscle.

LEARNED BEHAVIOR

Learned behavior is not innate. Organisms with complex nervous systems have more learned behavior. There are several different types of learned behavior. **Imprinting** is a type of learned behavior that occurs in very young animals. **Konrad Lorenz** first studied and described imprinting in his experiments with newly hatched geese. Lorenz was an ethologist. **Ethology** is the study of behavior in nature. Lorenz discovered that the young geese, or goslings, seemed to form social attachments to their mothers soon after hatching. This time of learning and forming attachments was called the **sensitive period.** It usually lasted only a few hours. Lorenz also discovered that if he separated the young geese from their mother and substituted another animal or object for the mother, the young geese would follow the "new" mother. Lorenz even became "mother" to a group of goslings.

Conditioning or **association** is another type of learned behavior. In conditioning, responses to one stimulus become associated with another stimulus. Conditioning was first described by **Ivan Pavlov,** a Russian psychologist, in 1900. Pavlov knew that dogs normally salivate when they see food. Pavlov's research showed that dogs could be conditioned to salivate when a bell was rung, even if no food was present. The bell became the primary stimulus instead of the food. Many of us have pets. We may use an electric can opener to open our pet's food at the same time we call out our pet's name. Soon the pet comes running when it hears the electric can opener. The pet now associates the sound of the can opener with food. Conditioning has occurred. If the pet is not given food after it responds to the can opener, the behavior will eventually stop.

Another learned, automatic response is habituation. A **habit** is a response that is repeated over and over until it becomes automatic. Habits can be broken, but it is very difficult to do so. Writing the year on the dateline of a check is a habit that must be changed with each new year. Writing your name, biting your fingernails, and brushing your teeth are other habits that become automatic after being repeated over and over again.

Trial-and-error learning occurs when an organism chances on a behavior that results in a reward. The organism then repeats the behavior hoping to receive another reward. Through practice the behavior becomes learned. If you were given a set of keys and told to open a door, you might have to try several keys before you discovered the correct key. If you repeated the process for several days, you would soon be able to use the correct key the first time without trying other keys in the lock. This is trial and error learning. In nature, young birds will peck at any small object. Through trial and error they eventually learn to peck only at edible objects.

Some animals use past experiences to help find solutions to new problems. This is called **reasoning** or **insight learning.** Because of their intelligence, primates are especially good at solving problems when placed in a difficult or unfamiliar situation.

Behaviors are necessary for survival. All organisms show some types of behavior. Research suggests that in complex animals, such as humans, there are fewer innate patterns and more learning involved in determining our behavior.

QUESTIONS

Multiple Choice

1. A response that is repeated so often that it becomes automatic is called   *a.* reasoning   *b.* conditioning   *c.* imprinting   *d.* habituation.

2. The scientist who first defined imprinting was   *a.* Ivan Pavlov   *b.* Konrad Lorenz   *c.* Charles Darwin   *d.* James Watson.

3. When birds build a nest it is an example of   *a.* instinct   *b.* learned behavior   *c.* a reflex   *d.* conditioning.

4. Which of the following behaviors is innate?   *a.* trial and error   *b.* conditioning   *c* .taxis   *d.* imprinting

5. Which of the following learned behaviors would be used in this situation? A chimp is placed in a room in which a bunch of bananas is hanging from the ceiling. Some boxes are scattered around the room. The chimp eventually figures out that by stacking the boxes and climbing on top of them she can reach the bananas.   *a*. imprinting *b*. reasoning   *c*. conditioning   *d*. habituation

Matching

6. innate
7. conditioning
8. imprinting
9. reflex
10. reasoning

a. uses past experiences
b. Konrad Lorenz
c. inborn
d. Ivan Pavlov
e. no thinking involved

Fill In

11. _____ neurons carry the impulse from the spinal cord to an effector.
12. _____ behavior is acquired during the lifetime of the organism.
13. _____ is a type of learned behavior that occurs in very young animals.
14. If an organism moves away from a stimulus, it is called a _____ response.
15. Reflexes occur in a certain pattern called a reflex _____.

Free Response

16. Explain how Pavlov determined that animals can be conditioned.
17. Give an example of a bad habit and tell how it could be eliminated.
18. Explain the path of a reflex arc and tell why it is an important behavior in animals.

19. Why is imprinting important for the survival of young animals?

20. What are some differences between innate and learned behavior?

Portfolio

Have you ever trained a dog or a cat? Write a guide for training that might prove useful to a person who has just adopted a pet. You might like to illustrate your training manual with drawings or photographs.

UNIT ELEVEN
ECOLOGY: WEBS OF LIFE

Chapter 39. The Biosphere and Biomes

Chapter 40. Ecosystems

Chapter 41. Humans and the
Environment

Chapter 39

The Biosphere and Biomes

ALL ORGANISMS ON Earth live in the relatively thin layer of soil, air, and water called the **biosphere.** Everything needed to support the wide variety of life on Earth is found in the biosphere. The energy to power the biosphere, however, comes from the sun.

ABIOTIC AND BIOTIC FACTORS

The biosphere contains abiotic and biotic factors. **Abiotic factors** are nonliving. They include things like soil, rain, sunlight, and temperature. Abiotic factors often interact with other abiotic factors. Abiotic factors also interact with, and affect, organisms in the biosphere. The living organisms in the biosphere make up the **biotic factors.** Plants, animals, protists, fungi, and monerans are all biotic factors. Like abiotic factors, biotic factors interact among themselves and also with the nonliving factors to create a stable, functioning system.

BIOMES

Scientists divide the biosphere into smaller areas called biomes. A **biome** is a geographical area that can be identified by characteristic biotic and abiotic features. **Terrestrial biomes** are land biomes. **Aquatic biomes** are water biomes. The abiotic

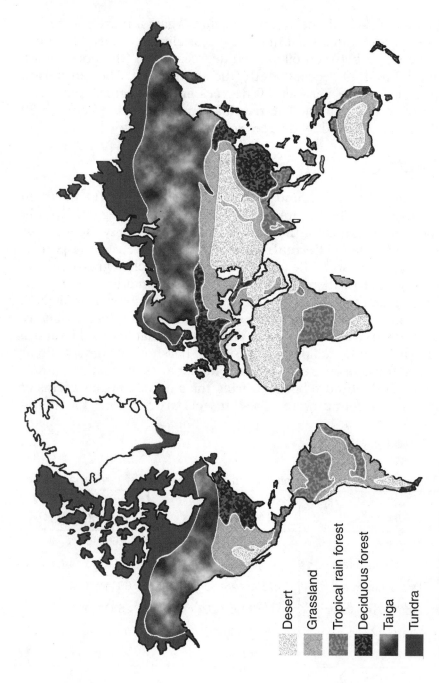

Figure 39-1. Biomes of the world.

Desert

Grassland

Tropical rain forest

Deciduous forest

Taiga

Tundra

factors of each biome determine what organisms are able to survive there. Biomes exist in three major climate zones. The **polar zone** exists between 60 and 90 degrees north (the north pole), and 60 and 90 degrees south (the south pole). The **temperate zone** is located between 30 degrees and 60 degrees north and south of the equator. The **tropical zone** is located between 30 degrees north and 30 degrees south of the equator.

Tundra

The **tundra** is found only in the polar zone of the northern hemisphere. The tundra is a very cold and dry region. Even in the summer, most of the soil remains frozen—only the topmost soil layer thaws. **Permafrost** is the name given to the permanently frozen soil. The tundra is a type of treeless grassland. No trees can grow there due to the permafrost and the short growing season that occurs when the air temperature warms slightly. Shrubs, mosses, and grasses are the predominant plants that are able to survive in this biome. The animals that live in the tundra also have to be adapted to this cold, harsh environment. Polar bears, musk oxen, caribou, and reindeer are some of the animals that call the tundra home. During the summer, huge swarms of insects are found there. These insects provide food for migratory birds.

Table 39-1. Average Precipitation and Temperature of Terrestrial Biomes

Biome	Yearly Precipitation (Average)	Yearly Temperature Range (Average)
Tundra	Less than 25 cm	−26°C to 4°C
Coniferous forest	35 to 75 cm	−10°C to 14°C
Deciduous forest	75 to 125 cm	6°C to 28°C
Tropical rain forest	More than 200 cm	25°C to 27°C
Grassland	25 to 75 cm	0°C to 25°C
Desert	Less than 25 cm	24°C to 34°C

Coniferous Forests

South of the tundra is a biome that has forests of cone-bearing trees such as spruce, pine, and fir. This biome is called the **coniferous forest** or the **taiga.** Like the tundra, the taiga has cold winters; but summers in the taiga are longer. This extra period of warm temperatures allows the ground to thaw and many plant species to flourish. Moose, bears, birds, and wolves are some of the animals that live in coniferous forests.

Deciduous Forests

Moving still farther from the north pole, the **deciduous forest** biome is found in the temperate zone, a climate zone characterized by long, warm summers and moderate rainfall. The deciduous forest biome is made up of oaks, maples, and beech

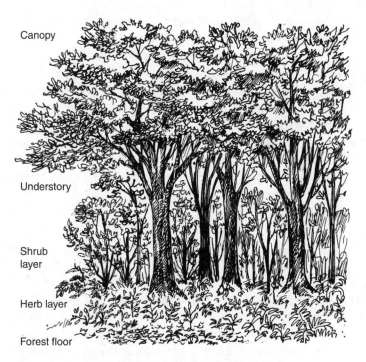

Figure 39-2. Stratification in a deciduous forest.

trees. These trees have broad, flat leaves, which they lose each autumn. **Vertical stratification** can be observed in this biome. The taller trees form the **canopy.** Smaller trees create an **understory.** Below the understory are the **shrub layer,** the **herb layer,** and the **forest floor.** This type of stratification, or layering, provides shelter, food, and protection for a variety of animals. In North America, deer, squirrels, insects, raccoons, birds, and rabbits live in deciduous forests.

Grasslands

A biome similar to the deciduous forest is the **grasslands.** This biome is the largest in North America. Grasslands are found in both temperate and tropical zones. Australia, Africa, Asia, North America, and South America all have grasslands. **Prairies, pampas, veldts,** and **steppes** are other names for grasslands. Trees do not grow very well in this biome due to the low rainfall, but many different types of grasses are well adapted to a grasslands climate. A **savanna** is a tropical grassland with a few scattered groups of trees. The best-known savanna, the Serengeti Plain, is located in Africa. Many burrowing animals, such as gophers and prairie dogs, live in the grasslands. Large herds of grazing animals also live in grassland biomes. In the United States, huge herds of bison once lived in the grasslands; today the herds are much smaller. In Africa, herds of zebras, gnus, and antelopes live in the grasslands. Large carnivores, such as lions, hyenas, leopards, and cheetahs, are also present.

Deserts

At latitudes of about 30 degrees north and south of the equator are the **desert** biomes. The largest desert on Earth is the Sahara. The deserts are the driest biomes. In some deserts, it rains only every few years. Plants are able to survive in a desert only if they are able to store water for long periods of time. **Succulents,** such as cactuses, are able to use their stems to store water. Many plants have waxy coatings on their stems and leaves to prevent water evaporation. Even though we think of deserts as being very hot, many have very cool nights. The animals and plants that live there must be able to adapt to a wide range of temperatures. Many of the animals that live in the desert are nocturnal. A **nocturnal** animal is active at night in order to avoid the high temperatures of the daylight hours. Birds, insects, scorpions, and reptiles such as snakes, tortoises, and lizards live in the desert.

Rain Forests

Most **rain forest** biomes are found in tropical regions near the equator. This biome has warm temperatures and large amounts of rainfall. Like the deciduous forest, the rain forest exhibits vertical stratification. There is one additional layer, the **emergent layer,** found in the rain forest. This layer is formed by the tops of the tallest trees in the forest. These trees are so tall that they reach above the canopy layer.

The soil of the rain forest is not very fertile. Most of the soluble nutrients are either used up quickly as plants grow or are washed away by the almost constant rain. The tropical rain forest biome contains the largest number of species of plants and animals in the world. Insects, birds, and monkeys are some of the animals found in this biome.

Marine Biomes

The ocean is a **marine biome.** Four oceans and many smaller seas cover about three-fourths of Earth's surface. The oceans can be divided into two main zones of life: pelagic and benthic. The **pelagic** region is the open ocean. Plankton are found in large numbers in the pelagic region. **Plankton** are usually microscopic organisms that are eaten by many species of fish that live in the open ocean. The **benthic** region is the ocean bottom. Many sessile animals are found in the benthic region. A **sessile** animal is attached to a surface, like sand or rock, and does not move from place to place. Sponges, sea anemones, corals, and clams are all sessile, benthic animals—at least for some stage of their life. The plants and animals that live in a marine biome are affected by many abiotic factors, including temperature, salinity, light, and pressure.

Freshwater Biomes

A **freshwater biome** is a body of water that lacks salt. Lakes, ponds, rivers, and streams are all freshwater biomes. Some freshwater biomes are made up of standing water. Lakes and ponds are standing water. Some are rich in vegetation and nutrients. These are called **eutrophic** lakes and ponds. Other lakes and ponds do not have a large amount of vegetation and are low in organic matter. These bodies of water are called **oligotrophic.** The water in other freshwater biomes is constantly moving. Rivers and streams are moving bodies of water. Fast-moving streams and rivers contain lots of oxygen due to the constant mixing of air and water. Slow-moving waters mix less and therefore contain lower levels of oxygen. The amount of oxygen in a body of water helps to determine what kinds of organisms can live there. Fish, birds, amphibians, and insects and other invertebrates live in and around freshwater environments.

The many biomes that make up the biosphere are different and unique. Each biome illustrates the diversity and adaptation of the millions of living things on Earth. Human activities can alter the characteristics of many biomes, resulting in problems for the organisms found there. It is important to understand the consequences of changing these geographical areas.

QUESTIONS

Multiple Choice

1. Which of the following biomes is found only in the polar zone of the northern hemisphere?   *a.* deciduous forest *b.* tundra   *c.* grassland   *d.* savanna

2. Animals that live attached to a surface are called *a.* abiotic   *b.* eutrophic   *c.* pelagic   *d.* sessile.

3. Another name for a coniferous forest is a   *a.* taiga *b.* tundra   *c.* pampa   *d.* canopy.

4. The biome with the least amount of rainfall is the
   a. grassland   *b.* rain forest   *c.* desert   *d.* deciduous forest.
5. All living things on Earth are found in the   *a.* abiotic
   environment   *b.* temperate zone   *c.* biosphere
   d. hydrosphere.

Matching

6. tundra	*a.* succulents found here
7. grassland	*b.* permafrost found here
8. marine	*c.* plankton found here
9. desert	*d.* taiga
10. coniferous forest	*e.* pampas, steppes

Fill In

11. In a deciduous forest, the small trees make up the _____.
12. If a lake is rich in vegetation and nutrients, it is termed
    _____.
13. The _____ region is the bottom of the ocean.
14. The _____ zone is located along the equator.
15. All of the nonliving factors in the biosphere are called
    _____ factors.

Free Response

16. How are desert organisms adapted to survive in a dry
    environment?
17. What are some factors that affect animals living in the
    ocean?
18. How is the savanna different from the normal grassland
    biome?
19. Why would it *not* be wise to destroy rain forests in an ef-
    fort to produce new farmland?
20. How would the slowing down of a fast-flowing river affect
    the animals living there?

Portfolio

People often try to balance the need to maintain a healthy environment against the need for industry to develop jobs and products to suit the needs of an ever-increasing human population. Pretend that you are a lawyer representing a company that wants to build a large factory on a parcel of land now covered by very old trees. What case would you present to build this factory?

Chapter 40

Ecosystems

PLAY A GAME of softball with friends and you know that interactions with other people occur frequently. These interactions are easily seen and sometimes felt, but interactions with other organisms and the environment are also occurring. For example, microorganisms that cause disease may be spread from person to person during the game. A dog may run across the field, causing the game to be halted. It may rain, and play might be suspended. The variety of possible interactions are almost endless.

ECOSYSTEMS AND POPULATIONS

The study of **ecology** examines the various relationships that exist between organisms and the environment. Ecologists may study interactions in large areas like the biosphere or a biome. They may also study interactions in smaller environmental units called ecosystems. An **ecosystem** is an area in which all parts of the environment, living and nonliving, interact and affect one another. Some ecosystems, like forests, are large; while others, like puddles of water, are small.

Ecosystems are made up of populations of different organisms. A **population** contains all members of a single species that live in a particular ecosystem. For example, all the deer living in a forest make up a population. The populations in an ecosystem make up a community. A **community** is all the plants, animals, fungi, protists, and monerans in a certain area. For example, the deer, rabbits, insects, plants, mushrooms, and all other life in a forest together form a community.

HABITATS AND NICHES

Populations tend to live in particular places in an ecosystem. Earthworms burrow in the ground, fish live in ponds, squirrels live in trees. The area in which an organism lives is called its **habitat.** In a habitat, every organism has a "job" or role. The role that an organism has in an ecosystem is called its **niche.** With each organism playing its role, the ecosystem functions smoothly.

Several different types of relationships may affect populations in an ecosystem. Zebras are grazing animals that eat green plants. Lions are **predators** that kill and eat zebras, their **prey.** This relationship holds peril for the zebras, which die in order for lions to survive.

Other relationships exist in nature. Several types of symbiotic relationships can occur between the different species in an ecosystem. **Symbiosis** is a close association between two different types of organisms. There are three major types of symbiosis. **Mutualism** is a relationship in which both organisms benefit. A lichen is an example of mutualism. Although it appears to be a single organism, a lichen is actually two different organisms, a fungus and an alga, that live together. Because it contains chlorophyll, the alga provides a source of food for the fungus. The fungus provides support and absorbs water and nutrients that the alga uses. Both the alga and the fungus benefit from this association.

Commensalism is a type of symbiosis that occurs when one organism benefits from the relationship and the other organism is neither helped nor hurt. Epiphytes are commensal organisms. **Epiphytes** are plants that grow on other plants. Some species of orchids are epiphytes. The orchid takes nothing from the tree it lives on except a place to perch. Some epiphytes in rain forests grow high on tall trees. This allows sunlight to reach the epiphyte. In the light, the epiphyte carries out photosynthesis. The tree that the epiphyte grows on is neither damaged nor helped by the epiphyte.

The third type of symbiotic relationship is called parasitism. **Parasitism** occurs when one organism benefits from the relationship while the other organism is harmed. The organism that benefits is the **parasite.** The organism that is harmed is called the **host.** An example of a parasitic relationship is a flea that lives on a dog. The flea receives its nourishment from the dog. The dog receives nothing from the flea. In fact, the dog loses blood and may develop a rash or an infection from the flea.

ENERGY FLOW

There is a definite flow of energy in an ecosystem. Almost all ecosystems on Earth ultimately derive their energy from the sun. Organisms that contain chlorophyll (plants are the most obvious example) are able to take the sun's energy and make their own food from it. Living things that have this ability are called **producers** or **autotrophs.** Algae and most plants are producers.

Other living things in ecosystems are unable to make their own food. They have to feed on other organisms to get energy. Organisms that eat other organisms are called **consumers** or **heterotrophs.**

If a consumer eats only plants, it is called an **herbivore** or a **primary consumer.** A cow is an herbivore. If a consumer eats only meat, it is called a **carnivore** or a **secondary consumer.** A tiger or a shark is a carnivore. If a consumer eats both plants and animals, it is called an **omnivore.** Humans are omnivores, although some people choose not to eat certain foods. Still other organisms in an ecosystem get their energy from feeding on dead organisms. These organisms are called **decomposers** or **saprophytes.** Fungi and bacteria are the major decomposers in ecosystems.

One way scientists show the flow of energy through an ecosystem is by drawing a food chain. A **food chain** shows the organisms that are involved in the flow of energy, starting with producers and ending with the decomposers. As an example, a food chain might begin with grasses (producers). The chain would continue with deer (primary consumers) feeding on the grass. Because they eat deer, wolves (secondary consumers) may be the next link in the chain. The chain would end with bacteria and fungi (decomposers) that break down the dead organisms into basic nutrients that become part of the soil. These nutrients are not wasted in the soil. They can be taken up by plants. Once again they become part of a food chain.

A single food chain does not represent everything going on in an ecosystem. In fact, there are numerous food chains in a single ecosystem. Many of these food chains overlap. A diagram that shows overlapping, interconnecting food chains is called a **food web.** Food webs provide a more accurate picture of feeding relationships and energy flow in an ecosystem.

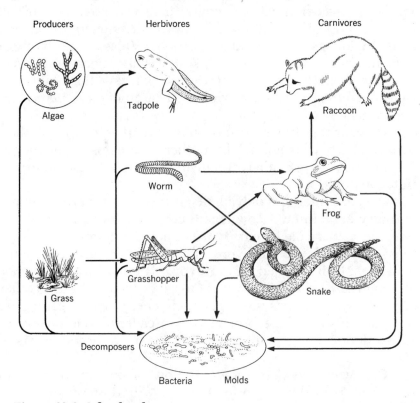

Figure 40-1. A food web.

PYRAMIDS

Several different types of ecological graphs can be drawn to show some of the relationships in an ecosystem. These graphs are usually drawn in a pyramid shape and so are called **ecological pyramids**. One type of pyramid is called a pyramid of biomass. A **pyramid of biomass** compares the total mass of the different types of organisms found in an ecosystem. Producers usually make up the base of the pyramid because they are the most numerous and have the greatest total mass. Primary consumers form the next level, and secondary and higher consumers form the next levels of the pyramid. Figure 40-2. illustrates the decreasing biomass in an ecosystem as you move from producers through the different levels of consumers.

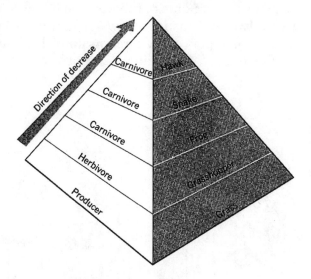

Figure 40-2. A pyramid of biomass.

A **pyramid of numbers** compares the number of organisms at each feeding level. A feeding level is also called a **trophic level.** Like the pyramid of biomass, the pyramid of numbers shows that producers usually outnumber consumers.

A **pyramid of energy** can also be drawn to compare the amount of energy in each feeding level. The amount of energy in the producer level is always highest. Energy is always lost as you move from one feeding level to the next. At each new feeding level, organisms use energy to maintain body functions, and therefore less energy is passed to the next higher level organism in the food chain.

NUTRIENT CYCLES

In ecosystems, the recycling of inorganic, or nonliving, materials enables substances that are in limited supply on Earth to be reused. Materials are reused in a series of nutrient cycles.

One very important cycle is the water cycle. The **water cycle** involves the movement of water from the air to the ground and back to the air. **Precipitation** is the movement of water from the air to the earth. Some of the water that falls to the earth becomes **groundwater.** This is freshwater that has soaked into the ground to become part of the streams and lakes that are under the Earth's surface. Most rainwater, however, moves along

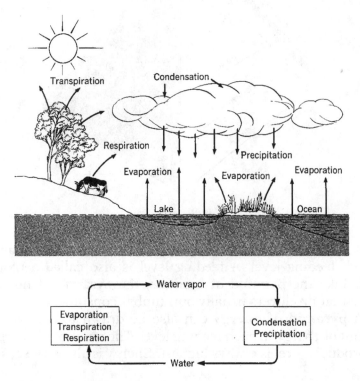

Figure 40-3. The water cycle.

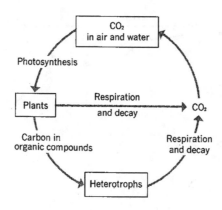

Figure 40-4. The carbon cycle.

the surface. It joins lakes, rivers, ponds, and streams. This is called **runoff** water. **Evaporation** carries water back into the atmosphere, where it condenses to form clouds. Later, the clouds may become so saturated with water that precipitation again occurs.

The **carbon cycle** is another important cycle in an ecosystem. In this cycle, carbon is passed between the carbon dioxide in the air and the organic molecules in living things. Both **photosynthesis** and **respiration** play important roles in the carbon cycle. During photosynthesis, plants use the carbon in carbon dioxide to form carbohydrates. When consumers eat plants, this carbon is passed from plant to animal. As organisms respire, or breathe, carbon dioxide is released into the atmosphere. Carbon dioxide is also released into the atmosphere when dead organisms decompose. Some carbon is trapped underground, where it becomes **fossil fuels,** such as coal and oil. When these fuels are burned, the carbon is again released into the atmosphere.

Nitrogen is an important part of living things. It is an element in protein molecules and in nucleic acids. More than 70 percent of the atmosphere is nitrogen gas, but most living things cannot use nitrogen in this form. In the **nitrogen cycle,** this gas is converted into **nitrates** and **nitrites** to be usable by organisms. Several chemical processes are part of the nitrogen cycle. **Nitrogen fixation** is a process in which **nitrogen-fixing bacteria** convert nitrogen gas into ammonium compounds. Many of these special nitrogen-fixing bacteria live in small swellings, called **nodules,** on the roots of plants such as peas, alfalfa, and

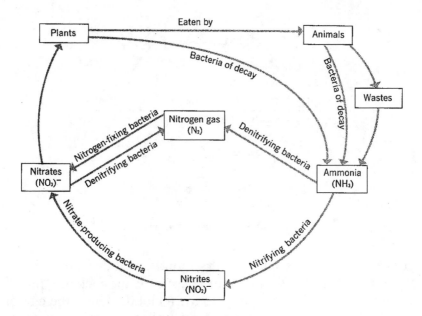

Figure 40-5. The nitrogen cycle.

soybeans. These plants are called **legumes.** Some plants can use ammonium compounds directly. Ammonium compounds in dead organisms can be produced by bacteria through a process called **ammonification.** Some plants cannot use ammonium compounds and need nitrogen that is converted into nitrates. This is carried out by **nitrifying bacteria** in a process called **nitrification. Denitrifying bacteria** are able to convert nitrates into nitrogen gas. There is a constant cycling of nitrogen from the atmosphere, through organisms, and finally back to the atmosphere.

The passage of oxygen through an ecosystem occurs in the **oxygen cycle.** Like the carbon cycle, the oxygen cycle requires the processes of photosynthesis and respiration. During photosynthesis, plants use energy from the sun to split water molecules. As a result, plants release oxygen into the atmosphere. This oxygen is used by most living things in the process of respiration. During respiration, carbon dioxide is released into the atmosphere. Energy and water are also produced. The water becomes part of the continuous water cycle.

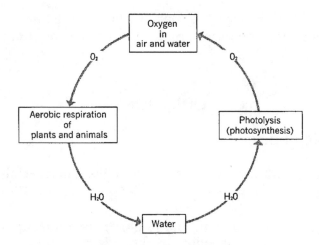

Figure 40-6. The oxygen cycle.

SUCCESSION

Many ecosystems undergo an orderly series of changes that is called **succession.** During succession, populations in an area are replaced by other populations. The first living thing to inhabit an area is called a **pioneer species.** In most land ecosystems, the pioneer species are lichens and grasses. Together with other organisms, a **pioneer community** begins to form. Over time, the grasses alter the soil and other conditions in an ecosystem. Small trees and shrubs may begin to appear in the community. Eventually a **climax community** appears. This last stage in succession might contain trees like maples and beeches. Because a climax community develops in response to local climate and geologic conditions, it tends to be stable. Its plants and animals are well adapted to survive in this particular area.

If succession occurs in an area that has not contained life before, like bare rock, the succession is called **primary succession.** If a community is destroyed, for example, by fire, succession begins to develop a new community. This is called **secondary succession.** Natural disasters, abandoned farmland, mining and logging can all produce a secondary succession.

Through the study of ecology, humans are better able to understand their role and relationships in the biosphere. The science of ecology clearly shows us that no organism exists by itself, separate from other organisms and the environment that surrounds it.

QUESTIONS

Multiple Choice

1. All the members of a given species living in a certain area make up a   *a.* community   *b.* biome   *c.* population *d.* habitat.

2. Which of the following is a symbiotic relationship in which both organisms benefit?   *a.* parasitism   *b.* mutualism *c.* commensalism   *d.* succession

3. Nitrogen-fixing bacteria are often found in nodules located on   *a.* epiphytes   *b.* climax communities   *c.* saprophytes *d.* legumes.

4. Succession occurring on a bare rock is   *a.* secondary succession   *b.* primary succession   *c.* tertiary succession *d.* a nutrient cycle.

5. Which of the following trophic levels would be found at the base of an energy pyramid?   *a.* decomposers *b.* consumers   *c.* herbivores   *d.* producers

Matching

6. niche   *a.* gradual change in an ecosystem

7. host   *b.* decomposer

8. symbiosis   *c.* job in an ecosystem

9. succession   *d.* close association

10. saprophyte   *e.* hurt by a parasite

Fill In

11. A consumer that eats meat is called a _____.

12. The movement of water from the air to the earth is called _____.

13. A _____ community is the last community in succession.
14. _____ is the study of relationships between living things and their environment.
15. A chart showing all the energy flow options in an ecosystem is called a food _____.

Free Response

16. Describe the succession that might occur after a forest is destroyed by a fire.
17. Draw a simple diagram of an ecosystem's water cycle.
18. Why is the role of decomposers important in an ecosystem?
19. Why is the sun important to an ecosystem?
20. Distinguish between a population and a community.

Portfolio

Draw a food web from your community. Include humans as part of the web. Compare your food web with the food webs prepared by others.

Chapter 41

Humans and the Environment

ALL ORGANISMS, including humans, depend on Earth for survival. Earth has a limited supply of materials needed to support life. Our supply of many natural resources is getting smaller and smaller as we use up many of the materials we need to live. Conservation is a way to help us protect those necessary materials. **Conservation** is the wise management of the Earth's natural resources.

RENEWABLE RESOURCES

Some of Earth's resources can be used and will be replaced in time. These are **renewable resources.** Wildlife, forests, and soil are all renewable resources.

Today, many wildlife species are threatened or endangered. In time, these animals may become extinct. **Extinction** occurs when a species disappears from Earth because no more individuals of that species are still alive.

Habitat destruction is a major cause of wildlife extinction. Laws protecting wildlife, breeding programs, and protected refuges are all ways to prevent, or at least slow, the rate at which some species are becoming extinct.

Forests are also an important renewable resource. However, in some areas forests are getting smaller because of an increased

**Figure 41-1. An artist's representation
of the extinct dodo.**

demand for wood and wood products. The needs of an ever-increasing human population have led to the deforestation of thousands of acres of forests. **Deforestation** occurs where large areas of forest are cut and cleared. In some places, tropical rain forests are cut and burned to clear land for farming. In a relatively few years, these cleared areas become unproductive when topsoil is washed away by the heavy rains that occur in these tropical areas.

After years of improper management, we now know that we must renew our forest resources by planting new trees to replace trees that were cut. The process of replacing cut trees is called **reforestation.** Reforestation will help insure that we have trees to harvest in the future and that wildlife will still have a habitat.

Soil is one of our most important natural resources. Good soil is needed to grow plants for food and for fibers to make cloth. Soil can be lost due to the actions of wind and water. Wind and water carry soil away in a process called **erosion.** Farmers have learned that they can limit the effects of erosion by using several different techniques. Some farmers use barriers of trees or shrubs to prevent wind erosion. These barriers are called **windbreaks.** The trees or shrubs interfere with the movement of wind and therefore limit the amount of soil blown away. Another method used to prevent erosion is **contour plowing.** In this technique, a farmer plows around, rather than up and down, a hill. Contour plowing helps prevent water erosion. On large hills, farmers may also plant crops on a series of terraces. **Terrace planting** is a process in which steps are carved into a hill. By eliminating part of the hill's slope, more soil remains in place.

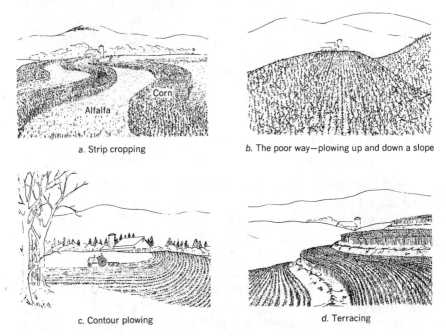

a. Strip cropping

b. The poor way—plowing up and down a slope

c. Contour plowing

d. Terracing

Figure 41-2. Ways to prevent soil erosion.

    Strip cropping is another farming method that is used to limit soil erosion. Here cover crops such as clover and alfalfa are planted in strips between bands of other crops. The roots of cover crops are adapted to hold soil in place. Farmers have also found that many fields lose nutrients after being used to grow the same crop year after year. This loss of soil fertility is called **soil depletion.** Farmers use crop rotation to prevent soil depletion from occurring. In **crop rotation,** farmers alternate the kind of plants grown on the same piece of land. By doing this, the same nutrients are not leached, or washed away, each year.

NONRENEWABLE RESOURCES

Other resources, such as water, coal, and mineral ores, are not replaced. These are called **nonrenewable resources.** Some of these resources would take millions of years to be replaced by natural processes. Water is one of our most important nonrenewable resources. All living things need water to survive. Today, many areas face water shortages, while other areas have an abundance of water. Farmers use water to irrigate their fields. Cities use huge amounts of water daily for drinking, bathing, recreation, and industry. In years with low rainfall, water may be rationed by governmental agencies. However, it would be a good idea to find ways to use less water even during times when water appears to be plentiful.

Watersheds must be protected. A **watershed** is land that eventually drains into a river system. Watersheds are the primary water source for New York City and other large cities. Some coastal cities get their freshwater by removing salt from ocean water. The removal of salt from ocean water is called **desalination.**

Coal, natural gas, and oil are another group of important nonrenewable resources. Because these fuels were formed from the remains of ancient organisms, they are called **fossil fuels.** Fossil fuels supply the energy needs of industrial societies. If we

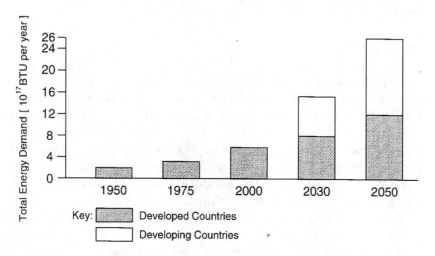

Figure 41-3. Past and projected world energy consumption.

were to continue to use only these fuels, we would eventually use up Earth's entire supply.

Today, we realize that we must conserve these fuels. We also have to develop ways to use alternative forms of energy and to develop new forms of alternative energy. We now use various forms of alternative energy. Many of these forms are much cheaper than fossil fuels. They also have the additional benefit of producing less pollution. Energy from the sun, **solar energy,** can be used to produce electric energy. Since solar energy captures the energy of the sun, it is never-ending and inexpensive.

We obtain **nuclear energy** by splitting atoms. When atoms are split, energy stored in the atoms' bonds is released. This energy can be used to produce steam to run turbines that produce electricity. The major problem with nuclear energy is finding safe ways to dispose of the wastes that remain. Some people are afraid of the possibility of an accident at a nuclear reactor, like the one that occurred at Chernobyl in 1986. At that time, large amounts of radiation were released into the atmosphere, contaminating air, land, and water over a large area.

Geothermal energy is another possible alternative to fossil fuels. **Geothermal energy** uses the heat formed deep within the Earth as an energy source. This heat can be used to produce steam to power turbines that generate electricity. **Wind energy** and **water energy** have also been used for many years to produce electricity.

Pollution

We release many materials into our environment that contaminate our surroundings. This is called **pollution.** The problem of pollution is most easily noticed in cities. **Air pollution** can be caused by many things, but most air pollution results from burning fossil fuels. Atmospheric conditions, like a temperature inversion, may combine with air pollutants to cause serious human health problems. A **temperature inversion** occurs when a layer of warm air becomes trapped between layers of cool air. This causes the air pollutants to become concentrated in the layer of cool air near the ground. These pollutants may remain trapped close to the ground for several days, creating dangerous health conditions.

Normal

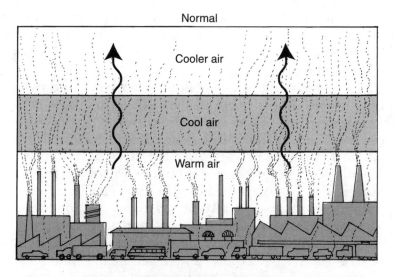

Temperature Inversion

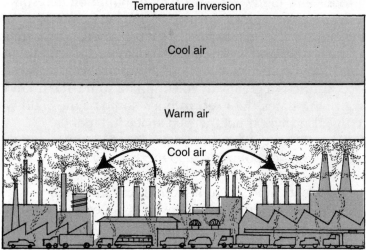

Figure 41-4. A temperature inversion.

In large cities, pollutants in the presence of sunlight create smog. **Smog** is a combination of smoke and fog. The chemical pollutants found in smog can vary from place to place. These chemicals can harm both plants and animals.

Acid rain is another very serious form of air pollution. Acid rain results from the burning of fossil fuels. Water in the air mixes with oxides produced when fossil fuels are burned. These oxides produce sulfuric and nitric acids when they combine with the water in the air. These acids fall to the earth in precipitation and damage many living organisms. Fish are especially susceptible to acid rain that falls on lakes and ponds. Trees are also damaged by acid rain.

Like air pollution, water pollution can be caused by many different things. Many pollutants are added to bodies of water from industries in the form of chemical waste products. Sometimes the pollutant is simply hot water from a factory. The addition of heat to a body of water is called **thermal pollution.** The increase in temperature can harm many organisms that live in the water. Many animals die because hot water holds less oxygen than does cool water. Water can also be polluted from **sewage systems, agricultural runoff,** and acid rain.

Land pollution occurs when people do not properly dispose of trash. Plastics, cans, bottles, and paper are all common sights along roads and highways. This type of pollution is responsible for destroying the natural beauty of our land as well as killing animals that eat it or become trapped in it. Unfortunately, proper waste disposal is now a problem as well. Landfills are becoming full as we dispose of more and more solid wastes each year. **Recycling** is a great way to eliminate some of our land pollution. It eliminates the need to burn some of the solid waste. Recycling also saves much-needed space in landfills.

Our need for resources increases as the human population increases. Levels of pollution also increase as the human population increases. In order to protect Earth for future generations, we must conserve our resources and find reasonable solutions to our pollution problems. It is hoped that advances in technology will help us solve the many environmental problems we face today and will face in the future.

QUESTIONS

Multiple Choice

1. The use of heat formed within the earth is called
   a. nuclear energy   *b.* geothermal energy   *c.* solar energy
   d. core energy.

2. Which of the following would prevent wind erosion?
   a. terracing   *b.* crop rotation   *c.* contour plowing
   d. windbreaks

3. The managing of our resources in order to protect them for
   future generations is called   *a.* ecology   *b.* conservation
   c. deforestation   *d.* recycling.

4. Thermal pollution occurs when hot water is added to a
   body of water. Which of the following results?   *a.* large
   amounts of oxygen are absorbed by the water   *b.* oxygen
   from the water is released into the atmosphere   *c.* animals
   become more active   *d.* animals live longer

5. Which of the following is not a renewable resource?
   a. forests   *b.* soil   *c.* wildlife   *d.* water

Matching

6. desalination
7. watershed
8. temperature inversion
9. reforestation
10. erosion

a. replacements for cut trees

b. drains into rivers

c. removing salt from ocean water

d. keeps air pollution near the ground

e. moving of soil

Fill In

11. _____ destruction is a major cause of the extinction of
    wildlife.

12. The alternating of cover crops with other crops to prevent
    erosion is called _____ cropping.

13. _____ is the clearing of large areas of forest.

14. _____ fuels supply most of our energy needs.

15. Many species have disappeared from Earth. We refer to
    these species as _____.

Free Response

16. Describe three methods of farming that control soil erosion.
17. Why is it important to conserve our forests?
18. How is acid rain formed in the atmosphere, and why is it dangerous to living things?
19. What are some ways you can help conserve Earth's supply of water?
20. Do you think nuclear energy is the best alternative to fossil fuels? Why or why not?

Portfolio

Little things can often make a big difference in the health of the environment. For example, suppose you and your friends each pick up one piece of paper on the way to and from school. Suppose you each use one glass of water less a day. Suppose you make one less trip a day in a car. Now suppose everyone in your town did the same thing. The impact would be enormous. Write and illustrate a small pamphlet using these and other similar ideas. The pamphlet could be handed out in your school and to your neighbors. The simple ideas you include in your pamphlet could have a major effect on the environment, and on you.

UNIT TWELVE
FUTURE HORIZONS
IN BIOLOGY

Chapter 42
Careers in Biology

THERE ARE MANY different and varied careers in the field of biology. Some careers require many years education, while others require much less education. Some biologists work indoors in laboratories or classrooms. Others work outdoors in fields or forests.

Because biologists study a variety of different topics, they use many different tools to measure, observe, and record data. Biologists interpret the data they collect to solve problems. This chapter outlines only a few of the many opportunities in biology you might want to explore as a career choice.

Biotechnologists apply a knowledge of molecular biology to create organisms that will produce things people need. Most biotechnologists use the science of genetics to help them change the genetic makeup of existing organisms to develop new or unique organisms. Bacteria are used for this purpose by most biotechnologists. Scientists are able to induce these organisms to produce important chemicals needed by the human body. Biotechnologists also develop strains of plants that show improved growth and crop yield. They even develop organisms that can help clean the environment. Biotechnology is a growing field of biology, with great potential for future jobs and discoveries. Biotechnologists need a strong background in genetics, microbiology, and biochemistry.

Health scientists enjoy working with people. **Physicians** and **nurses** take challenging courses in biology, chemistry, and

Figure 42-1. The study of biology offers many good career opportunities.

mathematics in college. Physicians continue their study in medical school for four additional years and then serve an internship and residency for several more years to receive specialized training. **Physical therapists** help improve a patient's ability to move. **Speech therapists** help individuals with speech problems.

Medical technologists are responsible for performing tests that are requested by physicians. For example, many different kinds of tests can be performed on a patient's blood. Physicians use the data collected by the medical technologist to help diagnose the patient's problem. Medical technologists must be able to use a wide variety of complex equipment such as centrifuges, automatic analyzers, and microscopes. Since the work of medical technologists is so closely linked to their equipment, these professionals must keep up with the constant changes that occur in the field of medical technology.

High school biology teachers provide students with information and laboratory experiences that illustrate various concepts in biology. Biology teachers give examples and conduct activities that make biology relevant to students' lives. These teachers try to help students develop a love of science. In order to make sure they cover the important points needed to understand a biological topic, biology teachers develop lesson plans.

After graduating from college, many biology teachers continue their learning through workshops. Some biology teachers also take additional courses at the college and postgraduate level. This continuing education keeps biology teachers aware of new developments in their field of science. Biology teachers receive state certification. Each state develops its own criteria for certification. College courses must include a good background in science, especially biology, as well as courses in education.

Ecologists usually conduct field research outside of a laboratory. Additional laboratory work may also be needed to test a particular finding or to analyze results obtained in the field. In their research, ecologists attempt to understand and explain how ecosystems are organized and function. One important concern for ecologists is to find ways to conserve natural resources. Ecologists seek to find solutions to environmental problems. Most ecologists spend hours collecting and analyzing data. Ecologists may work for state or local agencies, or private industries. Many work on special projects like designing parks to preserve wildlife. It is important for ecologists to be able to predict future problems and to decide what course we must take to avoid or deal with these problems.

Genetic counselors evaluate a couple's potential of having children with a genetic disorder. Many times a genetic counselor traces the history of a genetic disorder within a family. This helps the counselor determine the chances of the disorder occurring again. The counselor presents this kind of information to the couple in order to give them a better idea of possible options as they plan their family. The counselor must also be able to interpret prenatal testing. Most genetic counselors have backgrounds in genetics, microbiology, and psychology.

Marine biologists study life in the ocean. Many marine biologists specialize in certain kinds of organisms, others specialize in studying a certain part the ocean, still others study the relationships that exist among living things. Marine biologists may work for fishing industries, energy companies, or state and local governments. Most marine biologists have some background in biology, chemistry, and physics. Not all the work of a marine biologist is undertaken on the ocean. Many hours are spent in the lab making observations and recording data from collected samples.

Microbiologists study a variety of small, often microscopic organisms. Microbiologists usually study protozoa, bacteria, or fungi. Many microbiologists study disease-causing organisms. Others study organisms such as yeast, which are often helpful to humans. Microbiologists take courses in biology and chemistry, and are required to master a variety of microscopic techniques.

Biochemists explore the chemical processes that occur in living things. Biochemists help us understand how cells and the systems of the body actually function. Biochemists must have a strong background in chemistry, organic chemistry, molecular biology, and microbiology.

Botanists study plant life. Knowing how plants function is important to the future of our planet. The association between humans and plants is a very close one. We are dependent on the oxygen and food that plants make. Botanists study both the structure and functioning of plants.

The following list briefly sketches various additional careers related to biology. Read through them, go to your library or call a local college, and research any that you are interested in. Remember that scientific work is ever-changing, careers not dreamed of today may become reality in the future. You may find that you would enjoy a career in the biological sciences.

Entomologists study insects.
Zoologists study animals.
Ichthyologists study fish.
Herpetologists study reptiles.
Cytologists study cells.
Embryologists study the formation and development of the early stages of living things.
Paleontologists study prehistoric organisms.
Geneticists study inheritable traits and how they are passed from one generation to the next.
Mycologists study fungi.
Taxonomists classify living things.
Ornithologists study birds.
Ethologists study animal behavior.

Our knowledge of biology increases daily, and new questions constantly are being raised. In the future, new branches of biology will be formed, and new biologists will seek answers to unanswered questions. To be a biologist you must be willing to study what has already been discovered and use that information along with new research to find answers. Each career in biology provides an important link in the chain of biological knowledge.

QUESTIONS

Multiple Choice

1. Which of the following scientists would be needed to determine the importance of some of the chemical reactions occurring in the body?   *a.* botanist   *b.* microbiologist   *c.* ecologist   *d.* biochemist

2. Which of the following are the most common experimental organisms used by a biotechnologist?   *a.* fungi   *b.* protists   *c.* bacteria   *d.* animals

3. If you were in an accident and severely damaged your leg, which health worker would be trained to help you walk again?   *a.* medical technologist *b.* physical therapist   *c.* cytologist   *d.* respiratory therapist

4. Making sure we conserve our resources is the principal concern of   *a.* ecologists   *b.* biotechnologists   *c.* medical technologists   *d.* ethologists.

5. Which of the following professions also requires a strong background in psychology?   *a.* marine biologist   *b.* medical technologist   *c.* genetic counselor   *d.* ecologist

Matching

6. botanist   *a.* conservation of natual resources
7. zoologist   *b.* cells
8. cytologist   *c.* development
9. embryologist   *d.* plants
10. ecologist   *e.* animals

Fill In

11. While on a hike in the mountains, you see a bird you can't name. You take a photograph of it to help you remember its markings. When you return home, you take your photograph of the bird to an _____ to identify it.

12. A _____ studies the function and structure of plants.

13. A _____ would help a couple determine the chances of a disorder occurring in their unborn child.

14. Speech _____ help people who have problems with their speech patterns.

15. The scientists that study animal behavior are called _____.

Free Response

16. Why must you carefully plan the college courses you take in order to have a career in the biological sciences?

17. Where are some of the places biologists work? Give two examples.

18. Why would a biologist need to know about other areas of science, like chemistry, physics, or geology?

19. Name two types of biologists that might need to work together to solve a problem.

20. How will advances in technology affect many careers in the biological sciences?

Portfolio

Suppose you are invited to become the biologist on a spaceship setting out to explore a planet in another solar system. This planet may also support life. Write a fictional résumé that details the education and experiences that would qualify you for this position.

Chapter 43
The Future of Biology

ALMOST EVERY ASPECT of our lives is related in some way to biology. Without a thorough knowledge of biology, our world cannot be fully understood. Each day, biologists make discoveries that would have astounded their predecessors.

What types of questions will be asked of biologists in the future? We know from past experiences that new technologies will raise moral and ethical issues that need to be resolved. **Bioethics** is the term used when ethical concerns are involved in solving an issue in biology.

Many challenging questions involving bioethics are already being asked. In the future, biologists will have to resolve the ethical issues to come up with an appropriate solution to the biological questions.

ANIMALS AND RESEARCH

The use of animals in research and dissection is one important bioethical problem. In order to have a good experimental model to work with, many biologists feel that the use of experimental animals is necessary in research. Even though tissue cell cultures and computer technologies have provided alternatives to live animals in some cases, most scientists believe that these technologies cannot totally replace the use of animals in certain types of experiments.

Some people feel that animals should never be used in research. **Animal-rights activists** are very clear in their demands against the use of animals in research. In school situations, they would prefer to use videotapes of dissections rather than actual dissections, and computer models rather than live animals.

Currently, local and federal laws govern the use of research animals. These laws were primarily designed to control the pain that animals might experience during scientific research. Many laws also discourage the unnecessary use of animals in certain kinds of research. The issue of the use of animals for research will continue to be debated for many years to come.

ANIMALS AND ZOOS

A similar ethical debate involves animals in zoos. Unfortunately, some zoos do not maintain proper living conditions for the animals under their care. Some people view zoos as animal prisons. These people feel that the extinction of a species might be a better option than having animals live out their lives in zoos.

However, most people feel that zoos offer the best means of educating the public about the importance of saving endangered species. Zoos are also places where endangered species are bred so that young animals may be reintroduced to the wild. Some zoos even care for injured animals that are later returned to the wild.

The **captive-breeding programs** in some zoos have eliminated the need to capture certain animal species. Today, zoos trade certain zoo-bred animals with other zoos.

FUTURE AREAS OF BIOLOGICAL RESEARCH

It is doubtful that we will ever have answers to all the questions people pose about the natural world. In the future, new questions will be asked about the natural world. There may even be new branches of biology developed to better understand living things and the biosphere they inhabit. Throughout history, new technologies and discoveries have led to the formation of new branches of study. For example, the development of the aqualung in the 1940s by Jacques-Yves Cousteau opened up the ocean to exploration by biologists. The discovery of the structure of the DNA molecule by Watson and Crick in the 1950s expanded our understanding of the inheritance of traits. More recently, the development of new technologies in the 1980s and 1990s has given rise to the science of biotechnology. The future of biotechnology is certainly very bright.

Biotechnology uses living organisms and new techniques to produce things people need. One of the technologies involved in the science of biotechnology is **genetic engineering.** This technology gives scientists the ability to alter the genetic material in an organism by rearranging its genes or by combining its genes with those from other, unrelated organisms. Although it is carried by one organism, this new DNA has material from other organisms. The altered DNA is called **recombinant DNA.** Genetic engineers have been able to transfer human genetic material into bacteria, enabling the bacteria to produce two human hormones: insulin and human growth hormone. These bacteria are called **transgenic** because they contain functional recombinant DNA. The advantage of using transgenic bacteria is that large quantities of the needed substances can be produced by the rapidly dividing bacteria. Genetic engineering can also be applied to agriculture, medicine, and industrial products and processes.

When it is completed, the Human Genome Project will be a major boost to the science of biotechnology. (A genome is the complete set of chromosomes found in a species.) The Human Genome Project was begun in 1988. Since then, scientists have been attempting to identify the exact location of the thousands of gene pairs found on human chromosomes. This project is so complex that several laboratories have divided the work in order to complete the research in a reasonable amount of time. When it is completed, the Human Genome Project will provide scientists with the information needed to diagnose and treat many diseases.

Virology is another science that will experience increased scientific interest in the future. Virology is the study of viruses. Viruses are the cause of many serious diseases, including AIDS. Virologists of the future will be responsible for helping to answer the many questions regarding the structure and function of

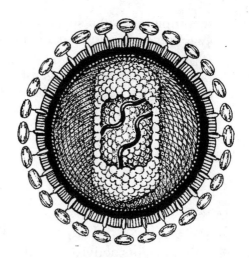

Figure 43-1. The AIDS virus.

viruses. They will provide important information needed to develop vaccines against viral diseases. Since viral particles are so small, the development of better microscopes and techniques will have a great effect on this area of science.

The history of biology has been filled with amazing discoveries, and the desire to find out more will insure that there are still many discoveries to be made. As teams of scientists work together to explore the living world, the twenty-first century will be a very exciting time for biologists.

QUESTIONS

Multiple Choice

1. The study of viruses is called   *a.* bacteriology   *b.* zoology   *c.* virology   *d.* protistology.

2. A combination of new DNA and old DNA from different organisms is called   *a.* a plasmid   *b.* recombinant DNA   *c.* replicated DNA   *d.* transcription.

3. Using organisms to produce things people need is called   *a.* ecology   *b.* microbiology   *c.* biochemistry   *d.* biotechnology.

4. A complete set of a species' chromosomes is called a
   a. genome   *b.* allele   *c.* phenotype   *d.* plasmid.

5. Which of the following has eliminated the need to capture
   large numbers of animals for zoos?   *a.* computer
   technologies   *b.* captive-breeding programs   *c.* dissections
   d. bioethics

Matching

6. videotapes

7. bacteria

8. virus

9. animal-rights activist

10. computer models

a. against use of animals

b. cause of AIDS

c. alternatives to dissection

d. common organisms used in
   genetic engineering

e. alternatives to animal
   research

Fill In

11. The Human _____ Project will provide scientists with the
    information needed to diagnose and treat diseases.

12. Some people even feel that the _____ of a species is a bet-
    ter option than having animals imprisoned in zoos.

13. _____ is the study of ethics applied as part of the solution
    to a biological issue.

14. Virologists of the future will provide important informa-
    tion needed to develop _____ against viral diseases.

15. One of the technologies involved in the science of biotech-
    nology is genetic _____.

Free Response

16. The Human Genome Project is costing a tremendous
    amount of money. Do you think the facts that will be
    learned from this research are worth the cost? Why?

17. Why will the science of virology be so important in the
    future?

18. Do you think zoos are the best way to save endangered species? Why?

19. Do you think animals should be used as dissection specimens in high school biology classes? Why?

20. What are some questions you think future biologists must answer in order for us to better understand living things?

Portfolio

In a brief letter to a friend, describe how the study of biology has changed the way you view the natural world.

INDEX